Assessing the Performance of Military Treatment Facilities

Nancy Nicosia, Barbara O. Wynn, John A. Romley

Prepared for the Office of the Secretary of Defense

Approved for public release; distribution unlimited

Center for Military Health Policy Research

A JOINT ENDEAVOR OF RAND HEALTH AND THE
RAND NATIONAL SECURITY RESEARCH DIVISION

The research described in this report was sponsored by the Assistant Secretary of Defense for Health Affairs and conducted by the RAND Center for Military Health Policy Research, a joint project of RAND Health, and the Forces and Resources Policy Center of the RAND National Defense Research Institute. The latter is a federally funded research and development center sponsored by the Office of the Secretary of Defense, the Joint Staff, the Unified Combatant Commands, the Department of the Navy, the Marine Corps, the defense agencies, and the defense Intelligence Community.

Library of Congress Cataloging-in-Publication Data

Assessing the performance of military treatment facilities / Nancy Nicosia ... [et. al].
    p. cm.
  Includes bibliographical references.
  ISBN 978-0-8330-4777-9 (pbk. : alk. paper)
  1. Medicine, Military—United States—Evaluation. 2. United States—Armed
Forces—Medical care—Evaluation. 3. United States--Armed Forces—Medical care—
Cost effectiveness. 4. Health planning—United States. I. Nicosia, Nancy.

  UH223.A92 2011
  355.7'20684—dc23

                                                              2011020359

The RAND Corporation is a nonprofit institution that helps improve policy and decisionmaking through research and analysis. RAND's publications do not necessarily reflect the opinions of its research clients and sponsors.

RAND® is a registered trademark.

Published 2011 by the RAND Corporation
1776 Main Street, P.O. Box 2138, Santa Monica, CA 90407-2138
1200 South Hayes Street, Arlington, VA 22202-5050
4570 Fifth Avenue, Suite 600, Pittsburgh, PA 15213-2665
RAND URL: http://www.rand.org/
To order RAND documents or to obtain additional information, contact
Distribution Services: Telephone: (310) 451-7002;
Fax: (310) 451-6915; Email: order@rand.org

Preface

This monograph considers potential efforts by the U.S. Department of Defense (DoD) to assess the performance of military treatment facilities (MTFs) in cost-effectively managing health care under DoD's TRICARE benefit. It offers an overview of performance assessment in the nonmilitary health care sector. It then analyzes the use of changes over time in average MTF utilization and costs as performance measures, focusing on how MTF size and catastrophic cases affect these metrics.

This research was sponsored by the Assistant Secretary of Defense for Health Affairs and conducted jointly by RAND Health's Center for Military Health Policy Research and the Forces and Resources Policy Center of the RAND National Defense Research Institute. The latter is a federally funded research and development center sponsored by the Office of the Secretary of Defense, the Joint Staff, the Unified Combatant Commands, the Department of the Navy, the Marine Corps, the defense agencies, and the defense Intelligence Community.

Comments are welcome and may be addressed to John Romley (romley@rand.org). For more information on the RAND Health Center for Military Policy Research, contact Susan Hosek at Susan_Hosek@rand.org or Terri Tanielian at Terri_Tanielian@rand.org. A profile of RAND Health, abstracts of its publications, and ordering information can be found at www.rand.org/health. For more information on RAND's Forces and Resources Policy Center, contact the Director, James Hosek. He can be reached by email at James_Hosek@rand.org; by phone at 310-393-0411, extension 7183; or by mail at the RAND Corporation, 1776 Main Street, Santa Monica, California 90407-2138. More information about RAND is available at www.rand.org.

Contents

Figures

Tables

Summary

The U.S. Department of Defense (DoD) has increasingly confronted financial, managerial, and operational challenges in sustaining the TRICARE health benefit, which it provided to 9.2 million beneficiaries in fiscal year (FY) 2006. Medical costs, for example, are projected to increase to 12 percent of DoD's total budget as of FY 2015, from a level of 8 percent in FY 2007.

In response to such challenges, the 2006 Quadrennial Defense Review motivated a transformation in business practices within the Military Health System (MHS). Performance-based planning and financing would allocate resources based on the value of activities to DoD's mission, while aligning accountability and authority within the system.

DoD has considered setting targets for health care utilization in its military treatment facilities (MTFs) and rewarding or penalizing MTFs according to their performance. Such an initiative supposes that MTF leaders are able to cost-effectively manage care, much as generalist physicians or managed-care plans are frequently expected to do in the private sector. For example, in areas in which TRICARE costs are high at private hospitals, MTF leaders may be able to encourage beneficiaries to be treated at military hospitals with spare capacity.

The Office of the Assistant Secretary of Defense for Health Affairs (OASD[HA]) has been monitoring utilization and costs "per member per month" (PMPM) among beneficiaries enrolled at each MTF in TRICARE Prime, a managed-care plan similar to a civilian health-maintenance organization. These PMPM metrics include all care received by beneficiaries, whether from the enrollment MTF, from other MTFs, or from civilian health care providers. OASD(HA) has considered assessing each MTF's performance by comparing current PMPM utilization with past levels.

Assessing changes in performance based on outcomes such as PMPM metrics raises a variety of important questions. What is the relationship between OASD(HA)'s metrics and MTF performance in cost-effectively managing care? What else may influence PMPM outcomes?

Figure S.1 suggests some answers. The figure shows OASD(HA)'s metric for inpatient utilization at DeWitt Army Community Hospital during FYs 2004–2005. Actual utilization in any quarter varies around the mean level. Performance may systematically influence mean utilization, yet there also appears to be some randomness.

Figure S.1 suggests some additional questions. If utilization were higher in FY 2006 than the FY 2004–2005 mean, how could OASD(HA) decide whether performance (or some other systematic

Figure S.1
Actual and Mean Inpatient Utilization at DeWitt Army Community Hospital, FYs 2004–2005

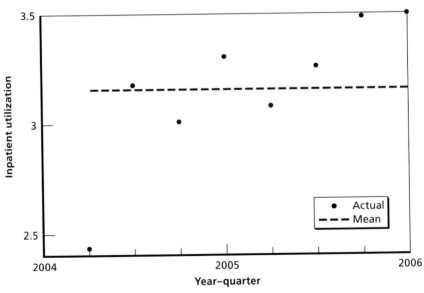

factor) had changed, or whether utilization just happened to be higher by chance? Is the nature of this decision concerning DeWitt, a relatively large MTF, similar to the decision that must be made at MTFs with small numbers of enrollees, where the randomness of utilization could be different? Do catastrophic cases, such as organ transplants, contribute to the random variability of inpatient utilization, making it harder to discern systematic changes?

Purpose and Approach

The purpose of this study is to help inform the sponsor's thinking about the assessment of MTF performance in general and the variability of MTF PMPM utilization and costs in particular. In broad terms, the study included a qualitative review of performance assessment in the nonmilitary health care sector, as well as a quantitative analysis of the variability of the sponsor's PMPM metrics and the roles played by MTF size and catastrophic cases.

For our qualitative review, we surveyed academic and policy research relating to performance assessment in health care. We visited a large Army hospital that served nearly 53,000 non-active duty Prime enrollees in FYs 2004–2005, where we interviewed MTF line administrators. We also conducted informal telephone interviews of experts in performance assessment at several private health care organizations.

This qualitative information helped guide the quantitative analyses, in which we were able to use two types of information:

- MTF-level data from FYs 2004 through 2006 on MHS-wide PMPM utilization and costs among TRICARE Prime beneficiaries enrolled at 114 "parent" facilities in the United States[1]
- disaggregate data for FY 2004 on admissions of Prime enrollees to military and civilian hospitals, as well as the personal characteristics of these beneficiaries.

---

[1]   As discussed below, some MTFs (such as small clinics) are "children" of "parent" facilities.

The analyses distinguished between inpatient, outpatient, and drug utilization. Active-duty personnel were excluded due to deployment-related data concerns.

We first analyzed MTF PMPM utilization and costs at both quarterly and annual frequencies. For each PMPM outcome at each MTF, we determined whether the change between FY 2006 and its mean level in FYs 2004–2005 was significant (as explained below). We then investigated the impact of an MTF's size on the variability of its PMPM outcomes and the frequency of significant changes. We defined size as the mean number of non-active duty enrollees during FYs 2004–2005; in some analyses, we considered five groups of similarly sized MTFs. We also considered the role of trends across MTFs in PMPM outcomes.

Separately, we analyzed the role of catastrophic cases in MTF performance assessment based on hospital admissions. We defined admissions as catastrophic if their diagnosis groups were typically associated with high levels of resource use. We then explored the role that catastrophic admissions played in PMPM inpatient utilization during FY 2004. We also simulated the impact of excluding these admissions on the identification of significant changes in noncatastrophic inpatient utilization during FY 2006.

Findings

Our qualitative review of performance assessment in the nonmilitary health care sector indicates that a variety of factors systematically affect health care outcomes, including PMPM utilization and costs, costs per provider or clinical episode, and so on. The performance of health care managers is such a factor. In our context, MTF leaders cause more or less care to be provided and care to be delivered more or less efficiently. Thus, MTF outcomes may be useful measures of performance assessment.

Health status is another systematic determinant of health care outcomes, since those who are less healthy typically need and use more care than others. Practitioners and researchers frequently attempt to account for health status by "risk adjusting" outcomes. Indeed, OASD(HA)'s PMPM metrics incorporate enrollee age, gender, and beneficiary status (e.g., retiree or dependent of a retiree). Such risk adjustments, while useful, are necessarily imperfect. When performance measures do not fully account for systematic factors, such as health risk or deployment of medical personnel, there can be substantial bias in assessments of MTF performance. The practical importance of this issue was beyond the scope of this study.

Utilization and costs also vary randomly. Whatever their health status, people use less care than usual in some periods and more in others. As a result, an observer cannot be certain about the true cause of a change in outcomes. On some occasions, an observer will mistakenly conclude that a change is systematic when in fact it is random ("false positives"). In other cases, the observer will conclude that a systematic change is random ("false negatives"). In reality, big changes are sometimes random noise, while small changes are sometimes meaningful.

An observer's confidence that a change is truly systematic can be enhanced by requiring that an outcome increase (or decrease) by a large magnitude. When this threshold is exceeded, an observed change is "statistically significant." A higher threshold for statistical significance results in fewer false positives, but more false negatives.

Given a confidence level, a lower rate of false negatives is desirable, because an observer has greater power to discern systematic changes. The false-negative rate is higher, however, when the randomness of an outcome is greater. PMPM utilization and costs may be more random at smaller MTFs, as there is less opportunity for enrollees' random health care needs to balance out when there are fewer enrollees. Catastrophic cases may also contribute substantially to the randomness of PMPM outcomes.

Table S.1 highlights some important findings concerning the frequency of statistical changes during FY 2006 when MTF outcomes are analyzed at a quarterly frequency. We found similar patterns (though generally higher frequencies) in the annual analysis. For outpatient utilization, drug utilization, and total cost, the frequency of significant changes was lower for the smallest MTFs than for the largest ones. For total cost, for example, the frequencies were 20.7 percent and 42.0 per-

Table S.1
Frequency of Statistically Significant Changes in FY 2006 from
FY 2004–2005 Mean Levels, Smallest MTFs Versus Largest MTFs

MTF Outcome	Smallest MTFs	Largest MTFs
Inpatient utilization	10.9%	3.4%
Outpatient utilization	19.6%	30.7%
Drug utilization	12.0%	15.9%
Total cost	20.7%	42.0%

NOTES: The smallest MTFs averaged no more than 7,187 non-active duty
enrollees during FYs 2004–2005; the largest MTFs averaged at least 27,911.
The confidence level is 95 percent.

cent. Changes in costs would be statistically significant in 5 percent of cases (given the 95 percent confidence level) even if there were no changes in the systematic determinants of outcomes. As a result, the share of significant changes in cost that are false positives could be as high as one in four (5%/20.7% = 24.2%) for the smallest MTFs, versus less than one in eight (5%/42.0% = 11.9%) for the largest ones. Unfortunately, the associated false-negative rates are unknown because the actual changes in performance and other systematic factors are unknown (though it would be possible to simulate these rates under various assumptions).

For inpatient utilization, the frequency of significant changes is actually lower at the largest MTFs. One possible explanation for this is that inpatient utilization became less variable at these MTFs. Among all MTF outcomes, the frequency of significant changes is lowest for inpatient utilization. While these outcomes were especially variable, the other outcomes tended to grow faster throughout the MHS in FY 2006, potentially making changes easier to discern. It is possible that such trends are partly attributable to changing performance across MTFs.

We also found that catastrophic cases, such as organ transplants and low-birthweight deliveries, play an outsized role in inpatient utilization. Diagnoses that ranked high in resource use accounted for a much larger share of utilization than of admissions. There is some reason to believe that excluding such cases would substantially increase the frequency of statistically significant changes in noncatastrophic inpatient utilization. It is possible, however, that MTF performance in managing catastrophic care is critical but hard to assess.

Altogether, our findings suggest that performance assessment of MTFs could be useful, though its effectiveness would generally be greater for larger facilities. Excluding catastrophic cases is practical and could be useful. In theory, systematic factors unrelated to performance could undermine the value of MTF outcomes as performance measures, and the practical importance of this issue may merit investigation. Finally, it is possible that alternatives, such as more targeted but complex assessments—for example, of cost per clinical episode—could help to diagnose MTF performance problems more reliably and to treat them more effectively.

Acknowledgments

We wish to acknowledge the assistance of MAJ Andrew Baxter, MAJ Mishaw Cuyler, and LTC Jean Jones, the programming of Stephanie Williamson, and the research assistance of Farrukh Suvankulov. We appreciate the support of project officers Robert Opsut and Pradeep Gidwani, as well as the assistance of Gregory Atkinson. John Adams and Kathleen Mullen provided valuable comments on this report; Dan Blum, Christine Eibner, Katherine Harris, Arvind Jain, and Jeanne Ringel provided helpful comments as this study progressed. Gordon Lee and Sydne Newberry assisted with communications; Nancy Good and Donna White helped with the document's formatting. We also wish to acknowledge the personnel at military, public, and private enterprises who participated in our interviews. Privacy concerns prohibit disclosure of their identities, but their generous assistance contributed greatly to this research.

Abbreviations

DMIS ID	Defense Medical Information System Identifier
DoD	U.S. Department of Defense
DRG	Diagnosis-Related Group
FY	fiscal year
GAO	U.S. Government Accountability Office
HCSR-I	Health Care Service Record–Institutional
MHS	Military Health System
MTF	military treatment facility
OASD(HA)	Office of the Assistant Secretary of Defense for Health Affairs
PMPM	per member per month
RVU	Relative Value Unit
RWP	Relative Weighted Product
SIDR	Standard Inpatient Data Record

CHAPTER ONE

Introduction

The mission of the Military Health System (MHS) is "to enhance the Department of Defense (DoD) and our Nation's security by providing health support for the full range of military operations and sustaining the health of all those entrusted to our care" (MHS, 2007). DoD has increasingly confronted financial, managerial, and operational challenges in achieving this mission, as health care costs have grown, the TRICARE health care benefit has expanded, and the MHS has supported military operations in Afghanistan and Iraq. The success of the MHS in meeting these challenges is important to defense policymakers, TRICARE beneficiaries, and the broader public.

In the face of these challenges, the 2006 Quadrennial Defense Review charted a "Roadmap for Medical Transformation" (TRICARE Management Activity, 2006) that includes transforming MHS business practices through performance-based planning and financing. The two components of business transformation are

1. a strategic, comprehensive, performance-based planning process for assessing goals and outcomes throughout the MHS
2. financial processes that allocate health resources based on the value of health care activities to the MHS mission and that align authority, accountability, and financial performance.

DoD-operated military treatment facilities (MTFs) may play a significant role in sustaining the TRICARE benefit through the efficient delivery of health care. DoD has considered allocating MHS funds to each service on the basis of the cost of efficiently delivered care (Opsut,

2006; TRICARE Management Activity, 2007b). Such an approach has been widely used in other settings to provide incentives to manage health costs. DoD has also considered setting utilization targets for MTFs and rewarding or penalizing MTFs for their performance with respect to the targets.

The Office of the Assistant Secretary of Defense for Health Affairs (OASD[HA]) has been monitoring utilization of TRICARE services and the associated costs "per member per month" (PMPM) among each MTF's beneficiaries who are enrolled in Prime, TRICARE's managed-care plan. We will refer to PMPM utilization and costs as MTF *outcomes* throughout this report. Changes over time in an MTF's outcomes can be used as measures of the performance of its leaders in cost-effectively managing care. For example, in areas in which inpatient care is relatively costly at private hospitals, MTF leaders may be able to encourage enrollees to be treated at military hospitals with spare capacity.

Utilization and PMPM costs also depend on factors largely beyond the control of their leaders. The health status of an MTF's enrollees systematically affects PMPM utilization and costs. In addition, there is random variation in health-related outcomes. Enrollees whose health is fundamentally stable use less care than usual in some periods but more in others. For example, a healthy person will have occasional checkups and accidental injuries.

For PMPM outcomes to be useful measures of MTF performance, random changes in these outcomes must be distinguished from changes that are systematic in nature, such as changed performance. An observer can have some degree of confidence that an observed change is not due to random variation but is, rather, "statistically significant." For a given change, the degree of confidence is greater when the random variability of the outcome is smaller.

Two factors are likely to affect the randomness of average MTF outcomes. First, PMPM utilization and costs may be highly variable at MTFs with small numbers of enrollees, because there is less opportunity for enrollees' health needs to "balance out" at small MTFs than at large ones. Second, catastrophic cases (such as organ transplants or low-birthweight deliveries) may contribute substantially to the variability of average MTF outcomes, because these cases are too resource-intensive to be balanced out by noncatastrophic care.

The goal of this study is to help inform the sponsor's thinking about MTF performance assessment in general and the variability of PMPM outcomes in particular. The study includes the following components:

- a qualitative review of existing approaches by researchers and practitioners to measuring performance within the nonmilitary health care sector
- an analysis of the variability over time of PMPM utilization and costs among MTFs' Prime enrollees, with emphases on the potential roles of
  - MTF size
  - catastrophic cases.

This monograph describes the results of the study and is organized as follows. Chapter Two summarizes the context for MTF performance assessment. Chapter Three offers an overview of performance assessment in nonmilitary health care and considers MTFs' PMPM outcomes from this perspective. Chapter Four uses OASD(HA)'s data to assess the variability of MTF outcomes and to identify significant changes in recent years, while focusing on the role of MTF size. Chapter Five explores the role of catastrophic cases with additional data on enrollees' hospital admissions. Chapter Six presents our conclusions.

CHAPTER TWO

The Context for Assessing the Performance of Military Treatment Facilities

In this chapter, we summarize the context for assessing the performance of MTFs. We characterize the mission of the MHS, explain the TRICARE benefit and its delivery, review the challenges confronting the MHS, and describe DoD's potential strategy of assessing MTF performance.

The Mission of the Military Health System

The mission of the MHS is "to enhance the Department of Defense (DoD) and our Nation's security by providing health support for the full range of military operations and sustaining the health of all those entrusted to our care" (MHS, 2007). This mission has three "pillars" (MHS, 2007):

- Provide a medically ready and protected force and medical protection for communities.
- Create a deployable medical capability that can go anywhere, anytime with flexibility, interoperability, and agility.
- Manage and deliver a superb health benefit.

The focus of this study is on the third pillar, although benefit management and delivery are related to the other pillars directly supporting military capability.

The TRICARE Benefit

TRICARE beneficiaries include active-duty personnel and their dependents, military retirees and their dependents, some reservists and their dependents, and some survivors. At the end of fiscal year (FY) 2006, approximately 9.2 million individuals were eligible for TRICARE (TRICARE Management Activity, 2007a).

The benefit is delivered at DoD-operated MTFs and by civilian health care providers. These two sources of care are known as direct and purchased care, respectively. In FYs 2004 and 2005, the direct-care system included 117 "parent" MTFs in the United States (Atkinson, 2007b).[1] These MTFs serve populations diverse in size and beneficiary mix within the context of local health systems that are themselves diverse, for example, in the availability of purchased care.

The TRICARE benefit covers inpatient and outpatient care, as well as drugs. A number of plans are offered for beneficiaries who are not Medicare-eligible, with terms (including fees and deductibles) that differ across plans. TRICARE Prime is a managed-care plan, similar to a civilian health-maintenance organization. TRICARE Standard is a fee-for-service plan that offers greater flexibility in choosing providers but also imposes higher out-of-pocket costs. Standard beneficiaries can pay lower cost shares by receiving care from a provider who belongs to the TRICARE Extra network. Our focus here is on TRICARE Prime beneficiaries, because OASD(HA) has been monitoring utilization and costs among each MTF's Prime enrollees.

TRICARE Prime beneficiaries are enrolled with a primary care manager, a health care provider who coordinates care, including referral to specialists within the direct- or purchased-care systems. More beneficiaries enroll in MTFs than in the civilian network (Mathematica Policy Research, 2007).

There is significant turnover in the total population of Prime enrollees at MTFs (Hanchate, McCall, and Ash, 2006). For exam-

---

[1]   MTFs may be "children" of a "parent" facility. For example, the Naval Branch Health Clinic at Naval Air Station North Island in Coronado, California, is child of Naval Medical Center San Diego. In a variety of settings including ours, data for child MTFs are "rolled up" to their parent facilities for reporting purposes.

ple, among 4.3 million Prime enrollees in FY 2002 for whom complete data were available, only 2.3 million were enrolled with the same primary care manager throughout FYs 2001 and 2002. We restrict our quantitative analysis to enrollees who are not active-duty personnel, because deployments to Afghanistan and Iraq may complicate the accurate measurement of utilization and costs for active-duty enrollees. The extent of turnover in active-duty dependents and retirees and their dependents is unclear. The extent of any turnover in the health status of MTF enrollees is also unclear but important, because health status affects an enrollee's typical health care needs and thus utilization and costs (see Chapter Three).

Active-duty personnel are required to enroll in Prime, and their dependents may do so with no fee or deductible. Retirees who are not Medicare-eligible may also choose to enroll for a fee of $230 for an individual or $460 for a family (TRICARE Management Activity, no date). There is no fee for those preferring to rely on TRICARE Standard/Extra. Prime enrollees receive priority over other beneficiaries for care at MTFs.

The benefit is jointly managed and implemented by several DoD organizations. OASD(HA) oversees military medical care. The TRICARE Management Activity, a DoD field activity charged with the benefit's administration, reports to OASD(HA). In addition, the Army, Navy, and Air Force have medical departments. These departments, each led by a Surgeon General, support the military pillars of the MHS mission while delivering the TRICARE benefit at their MTFs.

The delivery of care by MTFs is currently resourced through an activity group within the Defense Health Program's appropriation for operation and maintenance, as well as a separate appropriation for the labor costs of military personnel (U.S. Department of Defense, Task Force on the Future of Military Health Care, 2007).[2] The two lines of resources are allocated to the services for distribution to the MTFs and are not fungible for MTF leaders.

---

[2]   Another budget activity group funds base operations and communications relating to Defense Health Program facilities.

Challenges to the System

The MHS has confronted a variety of challenges in achieving its mission. In recent years, the system has increasingly come under financial stress. Between FYs 2000 and 2006, the unified medical budget increased from $17.4 billion to $38.5 billion (U.S. Department of Defense, Task Force on the Future of Military Health Care, 2007). The cost of medical care has escalated within the military, outpacing the increase in DoD's overall budget (U.S. Government Accountability Office [GAO], 2007b). In addition to inflation and cost-increasing improvements in medical technology, expansions in the benefit have contributed substantially to growth in total expenditures. TRICARE for Life, which provides medical care to Medicare-eligible retirees and their dependents, is an example. Medical costs were 8 percent of DoD's total budget in FY 2007 and are projected to be 12 percent of the total budget by 2015 if current trends continue (TRICARE Management Activity, 2007d).

The MHS has also been stressed operationally through its sustained support of military operations in Afghanistan and Iraq.[3] Combat units have needed medical attention in preparing to deploy, the returning wounded require care, and medical personnel have themselves been deployed.

DoD's Strategy

In the face of these challenges, the Quadrennial Defense Review has charted a "Roadmap for Medical Transformation" (MHS, 2007). This roadmap charts four activities: transforming the force, transforming the infrastructure, transforming the business, and sustaining the benefit. The relevant activity for this study—transforming the business—includes performance-based planning as well as performance-based financing.

---

[3]   The medical expenditures cited in the preceding paragraph do not include supplemental funding for the Global War on Terrorism in recent years. Global War on Terrorism funding for the MHS totaled $1.2 billion in FY 2006 (TRICARE Management Activity, 2007d).

Performance-based financing allocates health resources based on the value of activities to DoD's mission while aligning authority and accountability for financial performance. As an example, DoD has considered implementing a "prospective payment system" that would allocate resources to each service based on the cost of efficient care within the service's MTFs, instead of the cost of resources used to deliver care in the past (TRICARE Management Activity, 2007b). DoD has also considered setting utilization targets for MTFs and rewarding or penalizing MTFs for their performance with respect to these targets (TRICARE Management Activity, 2007b).[4] Both of these approaches suppose that MTF leaders are able to cost-effectively manage care, whether received from the enrollment MTF or elsewhere in the MHS, and therefore provide incentives to do so.[5]

Turning to performance-based planning, this element of business transformation calls for a strategic, comprehensive process for assessing goals and outcomes throughout the MHS.

DoD's Per Member Per Month Metric for MTFs

The performance of MTF leaders in cost-effectively managing care is relevant to business transformation, and so performance measures are needed. OASD(HA) has been monitoring each MTF's PMPM utilization and costs for its Prime enrollees (Atkinson, 2007a, 2007b; TRICARE Management Activity, 2007c). These metrics include all of the care received by an MTF's enrollees, whether from the enrollment MTF, from other MTFs, or from civilian health care providers.

---

[4] These initiatives also support the MHS's strategic plan (MHS, 2007). For example, they are relevant to the objective in the MHS "balanced scorecard" of managing DoD health care costs and shaping and sustaining the benefit. This objective contributes to the high-priority goals of sustaining the military health benefit through cost-effective, patient-centered care and effective long-term patient partnerships, as well as transformation to performance-based management for both force health protection and delivery of the health care benefit.

[5] Investments in strategic assets, such as information systems, are also relevant. Information technology can support process improvements, such as more efficient scheduling or staffing, that help MTFs to perform better in managing care in a cost-effective manner (MHS, 2007).

OASD(HA) has considered the performance of each MTF's leaders in cost-effectively managing care based on changes in PMPM utilization (TRICARE Management Activity, 2007b). The next chapter places these metrics in the context of outcome-based measures of performance in nonmilitary health care.

CHAPTER THREE

Performance Assessment in Health Care

In this chapter, we provide an overview of performance assessment in health care. This overview is based on our qualitative review of existing approaches, in which we surveyed academic and policy studies and interviewed leaders and analysts at several private health care organizations. We focus here on the example of Medicare costs by physician in order to help clarify the potential usefulness and limitations of performance assessment and, in doing so, to contextualize OASD(HA)'s metrics and what motivates our quantitative analyses of them.[1]

Outcomes and Performance

OASD(HA) has been monitoring outcomes, in particular, PMPM utilization and costs among Prime enrollees, at each MTF. In health care, education, and other settings, outcomes have regularly been used to assess various aspects of performance (Donabedian, 1988; Goldstein and Spiegelhalter, 1996). For example, hospital quality has been measured on the basis of mortality rates, while the efficiency of generalist physicians has been measured by total Medicare expenditures per patient (Office of Statewide Health Planning and Development, 2006; GAO, 2007a). Such measures, in order to be useful, must be influenced by performance.

---

[1] Linden et al., 2003, also provide a useful review of these issues.

The example of average Medicare costs by physician can help clarify. GAO analyzed the percentage of generalist physicians who cared for Medicare patients with high total medical costs (GAO, 2007a). This study was motivated by existing evidence that patients' overall care, and thus their total health care costs, are influenced by physicians, particularly generalists (see, e.g., Consumer-Purchaser Disclosure Project, 2003). GAO found that certain physicians cared for a higher-than-expected share of high-cost patients. For example, 20.9 percent of generalist physicians in Miami cared for such patients.

This finding suggests that a substantial number of physicians are "poor performers" in the costliness of their practice patterns. GAO concluded that the efficiency of the Medicare program could be improved by adopting a variety of practices that have been implemented by private health care payers, ranging from informing physicians about their performance relative to their peers to offering financial rewards to patients who choose efficient physicians. In reaching this conclusion, this outcome-based assessment of physician performance had to deal with several challenges to its reliability, which we consider in subsequent sections of this chapter.

In the DoD context, PMPM utilization and costs are likely to be influenced by the performance of MTF leaders in cost-effectively managing care under the TRICARE benefit. These leaders may cause more or less care to be provided, and cause care to be delivered more or less efficiently. For example, in areas in which inpatient care is relatively costly at private hospitals, MTF leaders may be more or less effective in having enrollees treated at military hospitals with spare capacity. Insofar as such drivers of performance are important but hard for OASD(HA) to monitor directly, PMPM outcomes may be useful measures of MTF cost-effectiveness. These outcomes may also be influenced by quality of care; DoD uses a different set of performance measures to monitor this aspect of performance (TRICARE Management Activity, 2007a).

Just as Medicare might vest responsibility with generalist physicians, so DoD might align accountability and authority in MTF leaders. Consistent with this notion (see Chapter Two), PMPM outcomes include all care that enrollees receive, whether from the enrollment MTF, from other MTFs, or from civilian health care providers.

DoD's situation is like that of a managed-care organization (indeed, TRICARE Prime is conceived as DoD's managed-care plan). These organizations sometimes contract with physician groups to care for beneficiaries, and PMPM utilization and costs are important metrics of performance in "covering lives" (McGlynn, Shekelle, et al., 2008). DoD also delivers care through its MTFs, and integrated delivery systems (such as Kaiser Permanente) must assess performance and provide incentives within their organizations.

Accounting for Other Determinants of Outcomes

In order to be useful, outcome-based measures must also be reliable indicators of performance. In reality, other factors also affect outcomes. Outcome-based measures of performance can be unreliable if these other factors are important influences. We will distinguish between systematic and random determinants of outcomes.

Systematic Factors

In health care, systematic factors include performance but also, at a minimum, the health status of the relevant population. In an analysis of average patient costs by physician, a doctor could be high-cost only because his or her patients are in relatively poor health, thus needing more care. Inefficient physicians can then be reliably identified only if costs are adjusted for the health risk of each physician's patients. The GAO study (2007a) did in fact adjust patient expenditures for health status based on a commercial technology also used to set payments for Medicare managed-care plans.

Such risk adjustment is widespread, although varied in implementation (see, e.g., Huang et al., 2005). In another example, the state of California reports risk-adjusted mortality rates among hospitals' patients, so that comparisons of hospital quality are not confounded by differences across hospitals in illness severity (Office of Statewide Health Planning and Development, 2006). In practice, risk adjustment is useful but nevertheless quite imprecise. For example, a person's medical spending in one year explains only 20–25 percent of the variation in spending in a subsequent year (Newhouse, Buntin, and Chapman, 1997).

Random Factors

There is also random variation in health care outcomes. A patient whose health is fundamentally stable nevertheless has some periods during which he or she receives some care and others in which he or she receives none; for example, a healthy patient will have occasional checkups and accidental injuries. In an analysis of average patient costs by physician, costs may differ among physicians, or for the same physician over time, because the health care needs of patients differ over time. Conclusions about physician efficiency (or MTF performance) must deal with this issue.

Given such randomness, an observer can never be totally certain that variation in outcomes is due to systematic factors, including performance. In practice, an observer must decide on the level of confidence that he requires in order to conclude that observed variation in outcomes is systematic in nature. In plain language, the question he must ask himself is, "How likely is it that this different outcome is not due to chance?" In reality, both randomness and systematic factors likely contribute to the observed changes. As a result, the minimum difference in outcomes that is required for the variation to be categorized as systematic follows from the "confidence level" and the outcome's random variability. If the observed difference (whether positive or negative) is larger than the required difference, the observer concludes that the variation is likely attributable to systematic factors. That is, in statistical terminology that we will use henceforward, the difference in outcomes is "statistically significant."

In such an exercise, an observer would occasionally make mistakes. In some instances, he would mistakenly conclude that the observed variation is statistically significant and thus systematic in nature, when in fact it is due to random chance. Such errors are known as "false positives." In other instances, he would mistakenly conclude that the variation is insignificant and thus random, when in fact it is due to systematic factors. These errors are known as "false negatives."

An observer can reduce the rate of false negatives by increasing the rate of false positives. For example, if 95 percent confidence were required, then the observer would encounter statistically significant variation in 5 percent of cases as a matter of chance, when there is in fact no systematic variation in outcomes. That is, the false-positive rate would be 5 percent. If the confidence level decreased to 90 percent, the required variation would shrink in magnitude, and the false-positive rate would increase to 10 percent. Truly systematic variation would then be more likely to exceed the larger threshold, and the rate of false negatives would decrease. An observer must make a tradeoff between the two types of errors based on his judgment about the harms from each.

For any confidence level (and thus for any false-positive rate), a lower rate of false negatives is desirable. An observer then has greater power to discern systematic variation in outcomes. As the randomness of an outcome grows, the rate of false negatives (given the confidence level) also grows. For this reason, random variability undermines the effectiveness of performance assessment.

Variability due to random factors tends to decrease with the number of patients (or enrollees) when costs or other outcomes are averaged. The reason is that the random components of patient outcomes tend to "balance out" during any period, with those experiencing relatively low outcomes offsetting those experiencing high ones. This phenomenon means that the minimum difference in outcomes required for statistical significance (given a confidence level) becomes smaller as the number of patients or enrollees grows. From a somewhat different yet consistent perspective, an observer's confidence that an observed difference is significant becomes larger with size. Appendix A provides a more formal discussion of statistical issues related to performance assessment.

In the GAO (2007a) study, the medical costs of a physician's patients (accounting for their health status) could be high either because the physician was inefficient or because his or her patients' health care needs turned out to be larger than average during the period studied.[2] GAO set thresholds for the proportion of physicians' patients who would be high-cost due to chance, based on the distribution of all patients' costs within each metro area studied. The GAO then concluded that some physicians in the area were inefficient because the percentage of physicians with high-cost patients exceeded this threshold.[3] Physicians with small Medicare practices were excluded from the analysis, because limited numbers of patients could lead to unreliable assessments. Based on a similar concern, mortality rates for California hospitals have not been reported for hospitals with relatively few patients (Office of State-wide Health Planning and Development, 2006).

Assessing MTF Performance Based on Outcomes

OASD(HA) has been monitoring each MTF's PMPM utilization and costs among Prime enrollees. Based on these metrics, it has contemplated whether to formally assess the performance of each MTF by comparing current performance with past levels, that is, by identifying changes in an MTF's performance over time. In the remainder of this chapter, we place this potential framework in the context of performance assessment in nonmilitary health care, while motivating the quantitative analyses of PMPM outcomes that follow.

Among health researchers and practitioners, identifying changes in the performance of providers or plans is less common than identifying differences in performance among them (GAO, 2007a; McGlynn, Shekelle, et al., 2008). Our interviews of representatives of several private health care organizations reinforced this judgment. Yet researchers have analyzed performance over time (see, e.g., Marshall, 1988; Bronskill et al., 2002).

---

[2]   Health status can also be an explanation insofar as risk adjustment of patient expenditures is imperfect.

[3]   Thresholds were set so that 1 percent of physicians would be expected to have an excessive proportion of high-cost patients, assuming that patients were equally likely to receive care from all physicians. In every area, some physicians were inefficient.

Identifying performance differences among MTFs could lead to different results than identifying performance changes at MTFs. Under the latter approach, MTFs might cycle over time between better and worse performance, if leaders believe that the likely benefits of improvements are more important than the costs of decreased performance.[4] These alternative benchmarks may raise additional issues of effectiveness and equity. Identifying performance changes at MTFs could result in large increases in performance among weak performers with substantial opportunities for improvement, and favorable assessments of these facilities; identifying performance differences among MTFs could result in relatively favorable assessments of historically strong performers. Each of these possibilities is desirable, though there is likely to be a tradeoff between them. Performance can be assessed according to both benchmarks simultaneously, as, for example, with the "value-based" purchasing program that Medicare has proposed for reimbursing hospitals (U.S. Department of Health and Human Services, Centers for Medicare and Medicaid Services, 2007).

Systematic Factors Other Than Performance

Whatever the benchmark, there can be incentives under outcome-based assessment to "game" the system based on factors other than performance. For example, hospitals may respond to "report cards" on health outcomes by restricting treatment of sicker patients (Dranove et al., 2003). Similarly, MTFs may be able to appear to improve their performance, without actually doing so, by limiting enrollment of sicker beneficiaries.[5]

---

[4]   We do not maintain that MTF leaders will behave in this fashion, only that such behavior is possible and potentially important. Understanding the response of MTF leaders to performance assessment is beyond the scope of this study but important. It may also be difficult. In any event, these responses are likely to be influenced by how DoD uses its performance assessments; this issue has not been decided.

[5]   If DoD were to compare performance among MTFs, each facility might be compared against its "peers," that is, facilities with similar characteristics (for example, service or size). It is possible that MTFs could alter some characteristics (for example, size) so as to be compared with a lower-performing peer group.

OASD(HA)'s potential framework deals with systematic factors other than performance in two ways. First, the utilization and costs of each member (that is, enrollee) are risk adjusted (TRICARE Management Activity, 2007c). OASD(HA) standardizes each enrollee into an "equivalent life" based on age, gender, and beneficiary category (e.g., active-duty family member). This approach, while practical and useful, is less sophisticated than other forms of risk adjustment, which are themselves imperfect (Hanchate, McCall, and Ash, 2006; Newhouse, Buntin, and Chapman, 1997).

Second, identifying changes over time may help purge PMPM outcomes of the influence of other systematic factors. Suppose, for example, that systematic factors include deployments of medical personnel as well as performance. Deployed personnel may have relatively efficient or inefficient practice patterns, or their absence may require that enrollees receive care from relatively low- or high-cost civilian providers, particularly in some important clinical specialties. If the number and characteristics of personnel who treat an MTF's enrollees are stable over time, then statistically significant changes in PMPM outcomes at an MTF must be due to changed performance.[6] This feature of OASD(HA)'s potential framework can be helpful when factors such as deployments are difficult to accurately monitor. A quantitative analysis of the effectiveness of these two methods for dealing with other systematic factors was beyond the scope of this study.

Random Variability in PMPM Outcomes

Random factors also lead to changes over time in MTFs' PMPM utilization and costs. As we discussed, enrollees' health care needs during any period are inherently random. In addition, instability in systematic factors that OASD(HA) is unable to monitor contributes to variation in outcomes; from OASD(HA)'s perspective, this variation is also random. Deployment of medical personnel is a potential example. If deployment is not stable over time, and if deployment is independent of MTF managers' performance given the resources available, statistically significant changes in PMPM outcomes must still be due to changed

---

[6]   We consider the possibility that other systematic factors are unstable momentarily.

performance. OASD(HA) would, however, require a larger change in outcomes for significance, given its desired confidence level. If deployment were not independent of performance, then significant changes could be due to deployment rather than performance; furthermore, deployment could obscure changes in performance.

This random variability is central to the quantitative analyses that follow. PMPM utilization and costs are averaged over the number of equivalent lives enrolled at each MTF. As we explain in the next chapter, the average number of Prime enrollees (excluding active-duty personnel) ranged from 824 at the smallest parent MTF to 85,109 at the largest parent MTF during FYs 2004 and 2005. As noted earlier in this chapter, the random variability of PMPM outcomes is likely to be greater at smaller MTFs. We therefore analyze in Chapter Four the impact of MTF size on the variability of PMPM outcomes, the magnitude of changes required for statistical significance, and the frequency with which observed changes are significant.

To be clear, we do not focus on the impact of MTF size on the *levels* of PMPM outcomes. This "volume-outcome" relationship may be important in our context, as it is in others (see, for example, Luft, Hunt, and Maerki, 1987). The scope of the present study, however, is to compare each MTF's current performance with its past performance; MTF size is only relevant due to its potential impact on the variability of outcomes.

Catastrophic cases, such as organ transplants and low-birthweight deliveries, also contribute to the random variability of PMPM outcomes. As a result, these resource-intensive cases may make the identification of systematic changes in outcomes more difficult. Appendix A provides numerical examples of the potential impacts of catastrophic cases and MTF size on performance assessment. In Chapter Five, we explore the actual role of catastrophic cases in PMPM inpatient utilization. Excluding catastrophic cases may be useful, yet the practicality of doing so using MHS data systems is uncertain. We also explore this issue.

PMPM Outcomes Versus Episode-Based Performance

Assessing MTF PMPM utilization and costs contrasts with another broad framework for performance assessment in health care. Under this alternative, the cost or quality of "episodes" of care is assessed (GAO, 2007a). These episodes bundle together a "block of one or more medical services received by an individual during a period of relatively continuous contact with one or more providers of service, in relation to a particular medical problem or situation" (Solon et al., 1967). A catastrophic case at a hospital is not conceptually equivalent to an episode of care, because an episode may include multiple encounters with multiple health care providers.

While our quantitative analyses were limited in scope to PMPM outcomes, the advantages and disadvantages of episode-based performance assessment nevertheless merit discussion. Administrative complexity is a disadvantage of the episode-based approach. Data on care delivered by different providers on different occasions, if available, must be accurately linked into episodes.

Episode-based performance assessment has other disadvantages. First, the utilization of alternative treatments may not be measured accurately, depending on how episodes are defined. For example, the beginning of a new episode of a recurring problem may be identified by a doctor's visit with the relevant diagnosis, even though the patient had recently been taking medication for the problem. Second, effective prevention strategies reduce the need for treatment, potentially lowering costs and improving health. This aspect of performance may not be captured by an analysis of treatment episodes; PMPM costs do reflect all aspects of performance.

An advantage of episode-based performance assessment is that the results may be relatively "actionable." For example, evidence that costly and invasive procedures are used for coronary care where drug therapy is appropriate could help in diagnosing and treating an MTF's "cost problem." Yet OASD(HA) would not need such information if MTF leaders had access to equivalent information and appropriate incentives to use it.

Another potential advantage is that episodes are defined for specific medical conditions. Thus, differences in the mix of beneficiaries' medical conditions would not systematically affect MTF performance assessments, making this approach relatively reliable. Furthermore, the occurrence of episodes is somewhat random. This randomness does not affect performance assessments based on actual episodes. Under this approach, then, it may be possible to identify more statistically significant differences in performance. It is also possible, however, that MTF leaders are able to manage the incidence of some medical conditions or the occurrence of certain episodes of care. If so, DoD might wish to assess performance in these dimensions but could not use the episode-based approach to do so.

CHAPTER FOUR

Performance Assessment and MTF Size

This chapter explores changes in MTF outcomes over time based on OASD(HA)'s metrics for FYs 2004–2006, focusing in particular on the role of MTF size. We first describe the data set and then the analysis sample and outcomes. After characterizing MTF size, we describe the relationship between outcomes and size. We then assess changes in MTF outcomes over time. We also consider some additional issues, including the confidence level for statistically significant changes, performance assessment of Services as a whole, the length of the assessment period, and assessment relative to MHS trends. Chapter Five will explore the role that catastrophic cases play in assessing MTFs based on additional data.

Data Set on MTF Outcomes

OASD(HA) shared its data set on PMPM outcomes from the first quarter of FY 2004 through the second quarter of FY 2007 (Atkinson, 2007a, 2007b).

This data set includes the numbers of Prime enrollees and equivalent lives, as well as total utilization and costs among Prime enrollees, for parent MTFs on a monthly basis. Each of these fields is reported by beneficiary class, including active-duty personnel, the dependents of active-duty personnel, and retirees and their dependents.

Utilization is disaggregated into inpatient care, outpatient care, and drugs. Inpatient utilization is based on Relative Weighted Products (RWPs), a DoD measure of workload that represents the relative resource consumption of patients' hospitalizations (DoD, 2002). Outpatient utilization is based on Relative Value Units (RVUs), a valuation or rating of physician services on the basis of relative physician resource inputs to provide medical services (TRICARE Management Activity, 2002). Drug utilization is measured by the number of 30-day-equivalent prescriptions filled.

These measures are used in both military and nonmilitary health care settings to measure resource use across varied clinical circumstances. For example, if two MTFs provide a hospital stay where one is for pneumonia while the other is for an organ transplant, the RWP measure of inpatient utilization will be higher (all else equal) at the latter MTF, as is appropriate.

Utilization and costs "roll up" (that is, include) all care received by each MTF's enrollees, whether from the enrollment MTF, from other MTFs, or from civilian health care providers.

Analysis Sample and Outcomes

We analyze 114 parent MTFs located in the United States that appeared in the data set in every month.[1] A complete list of these MTFs can be found in Appendix B.

Table 4.1 summarizes the outcomes we analyze, which include inpatient, outpatient, and drug utilization. OASD(HA)'s thinking about MTF performance assessment has focused on utilization (Opsut, 2006). We also consider total costs, as these data were readily available.

OASD(HA) currently monitors monthly inpatient utilization (RWPs) per 1,000 equivalent lives, monthly outpatient utilization (RVUs) per 100 equivalent lives, monthly scripts (30-day-equivalents) per 100 equivalent lives, and annual total cost per equivalent life. As explained in Chapter Three, equivalent lives standardize enrollees according to their health risk.

---

[1]   Three MTFs did not appear in the data during at least one month.

Table 4.1
MTF Outcomes Among Non-Active Duty TRICARE Prime Enrollees,
FYs 2004–2005

	Monthly Inpatient Utilization per 1,000 Equivalent Lives	Monthly Outpatient Utilization per 100 Equivalent Lives	Monthly Scripts per 100 Equivalent Lives	Annual Total Cost per Equivalent Life
Mean	6.15	64.1	99.3	$2,351
Minimum	2.07	46.5	65.8	$1,709
Maximum	26.37	114.9	143.4	$6,904

NOTE: Outcomes are measured at quarterly frequency and averaged over
FYs 2004–2005.

In our benchmark analysis, we use quarterly averages of these monthly outcomes. We also analyze annual averages later in this chapter. There is a potential tradeoff between a higher frequency of assessment and greater confidence in the results due to less variability in the assessed outcomes.

MTF Size

As we explained in the last chapter, MTF size may affect the variability of PMPM outcomes and thus the frequency with which statistically significant changes can be identified.

Table 4.2 summarizes mean enrollment and equivalent lives at these MTFs during FYs 2004–2005. We restrict our analysis to the dependents of active-duty personnel and retirees and their dependents, because OASD(HA) raised concerns about measurement error for active-duty personnel arising from deployments in support of military operations in Afghanistan and Iraq. The table therefore distinguishes between total and non-active duty enrollment and equivalent lives.

MTFs differ greatly in their size during FYs 2004–2005. In our analyses, we will follow Hanchate, McCall, and Ash (2006) in defining MTF size by the number of Prime enrollees. Because we exclude active-duty personnel from our quantitative analyses, we use the number of

Table 4.2
MTF Prime Enrollment and Equivalent Lives, FYs 2004–2005

Statistic	Enrollment		Equivalent Lives	
	Total	Non-Active Duty	Total	Non-Active Duty
Mean	26,728	18,367	27,219	19,792
Minimum	3,837	824	3,989	821
Maximum	121,079	85,109	116,328	89,339

NOTE: Outcomes are measured at quarterly frequency and averaged over
FYs 2004–2005.

non-active-duty enrollees.[2] MTF size ranged from a minimum of 824 enrollees to a maximum of 85,109 enrollees, that is, by a factor of more than 100.

In much of our analysis, we will characterize size according to five groups of similarly sized MTFs. The first group includes the smallest 20 percent of MTFs, which we refer to as the first (or smallest) size "quintile." The second quintile includes the next smallest 20 percent of MTFs. The third and fourth quintiles are defined similarly. The fifth and largest quintile includes the largest 20 percent of MTFs. Table 4.3 reports the size range associated with each quintile; Appendix Table B.1 lists MTFs in order of non-active duty enrollment.

The Relationship Between the Variability of MTF Outcomes and Their Size

With data on outcomes throughout FYs 2004–2006, we can identify statistically significant changes in each MTF's outcomes in 2006 from their 2004–2005 means. Such changes are deemed to be systematic (i.e., nonrandom) in nature and potentially due to changed performance. We estimate the random variability of outcomes based on the data for FYs 2004–2005.

---

[2]   The relative sizes of MTFs are very similar under the alternative possible measures in Table 4.2, with correlation coefficients with the number of total enrollees in excess of 0.97.

Table 4.3
Range of MTF Sizes in Each Size Quintile

Size Quintile	Size Range
1 (smallest)	824–7,187
2	7,403–10,990
3	11,030–16,890
4	17,569–27,828
5 (largest)	27,911–85,109

NOTE: Size quintiles are defined by mean non-active duty Prime enrollment in FYs 2004–2005.

Figure 4.1 illustrates inpatient utilization at DeWitt Army Community Hospital during the eight quarters in FYs 2004–2005. The mean level during this period (which corresponds to the horizontal line in the figure) was 3.16 RWPs per 1,000 equivalent lives. Utilization in each quarter (which corresponds to the dots in the figure) varied randomly—and substantially—around this mean. Thus, as Chapter Three explained, an observer could not have had complete confidence that a change in utilization in 2006 from its 2004–2005 mean was statistically significant, that is, deemed to be systematic in nature.

We will characterize the variability of each outcome by its standard deviation, a statistical measure defined and explained in Appendix A. To begin with, we cannot, as a practical matter, illustrate the variability of each outcome at each of the more than 100 MTFs analyzed with its own figure. More importantly, the standard deviation appears in the formula that determines statistical significance.

Figures 4.2–4.5 illustrate the relationship between size and the standard deviations of inpatient utilization, outpatient utilization, drug utilization, and total cost among the MTFs analyzed.[3] In these figures, dots correspond to MTFs, while a line shows the overall trend. For

---

[3]   We estimate MTF-specific means and standard deviations using the eight quarters of available data in order to obtain a reliable estimate for each. It is possible that there are trends in both the mean and standard deviations for each MTF that would be observable if these calculations were updated, for example, over the most recent eight quarters.

Figure 4.1
**Actual and Mean Inpatient Utilization at DeWitt Army Community
Hospital, FYs 2004–2005**

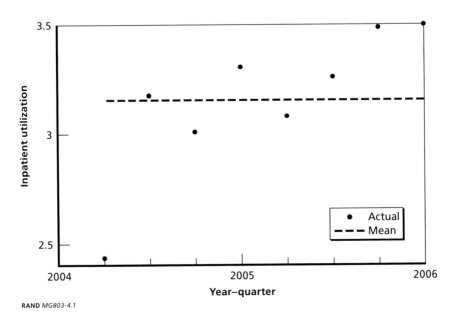

RAND *MG803-4.1*

each outcome, standard deviations tend to decrease with MTF size. This pattern is consistent with the hypothesis from Chapter Three that there is greater opportunity for enrollees' health care needs to balance out at MTFs with relatively large numbers of enrollees.

The variability of an outcome might tend to shrink or grow with its mean level. The seemingly small standard deviations at large MTFs in Figures 4.2–4.5 could then be large in comparison with mean outcomes, if their mean levels strongly decreased with MTF size. Figures 4.6–4.9 illustrate the relationship between MTF size and mean outcomes. Mean outpatient utilization tended to decrease slightly with MTF size. For the other outcomes, mean levels increased with size. These trends could be attributable to size-associated differences in the efficiency of care, to unmeasured heterogeneity in enrollee health status, or to measurement error. At the sponsor's request, we explore the size relationship for outpatient utilization in Appendix C.

Figure 4.2
Standard Deviations of Inpatient Utilization and MTF Size, FYs 2004–2005

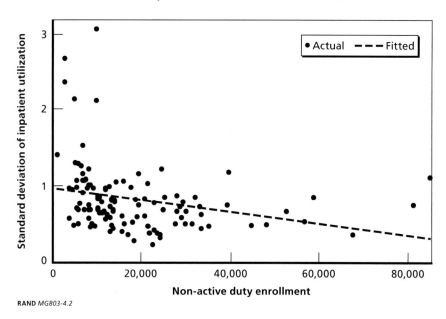

Figure 4.3
Standard Deviations of Outpatient Utilization and MTF Size, FYs 2004–2005

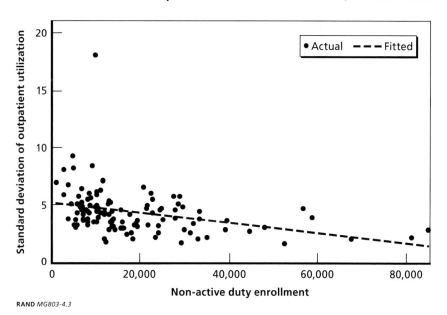

Figure 4.4
Standard Deviations of Drug Utilization and MTF Size, FYs 2004–2005

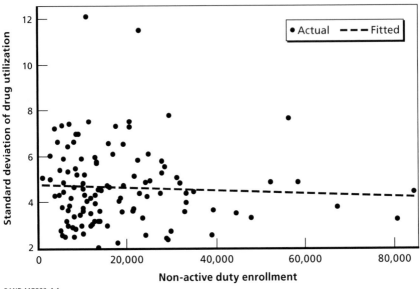

RAND *MG803-4.4*

Figure 4.5
Standard Deviations of Total Cost and MTF Size, FYs 2004–2005

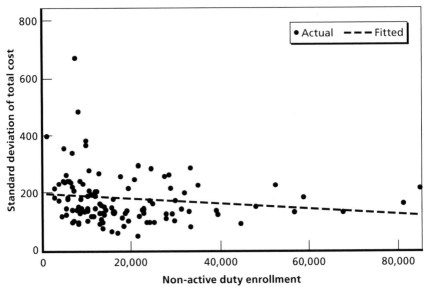

RAND *MG803-4.5*

Figure 4.6
Mean MTF Inpatient Utilization and Size, FYs 2004–2005

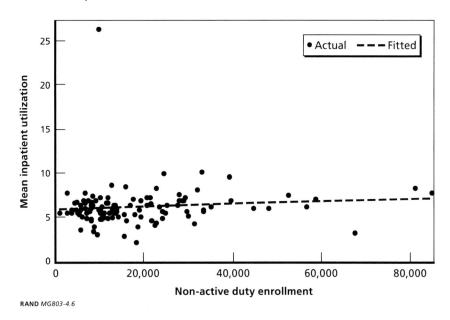

Figure 4.7
Mean MTF Outpatient Utilization and Size, FYs 2004–2005

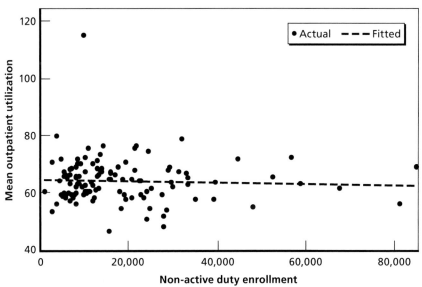

Figure 4.8
Mean MTF Drug Utilization and Size, FYs 2004–2005

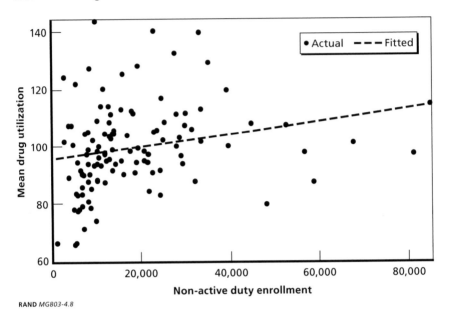

RAND *MG803-4.8*

Figure 4.9
Mean MTF Total Cost and Size, FYs 2004–2005

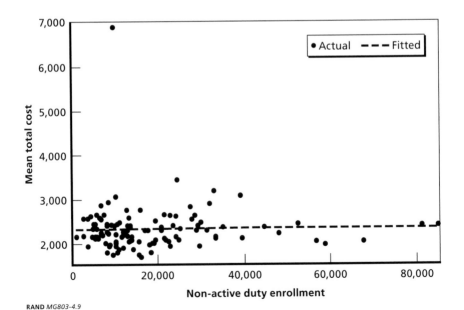

RAND *MG803-4.9*

The explanation for the trends in the other outcomes, while potentially important, was beyond the scope of this study.

Assessing Changes in MTF Outcomes

We are able to formally assess the changes in outcomes observed over the four quarters of 2006.

For each outcome at each MTF, we determine the magnitude of the change required during FY 2006 for statistical significance with 95 percent confidence based on its estimated mean and standard deviation during FYs 2004–2005. These changes are defined in percentage terms, because mean outcomes tended to vary with MTF size (see Figures 4.6–4.9).

The required changes are described in Table 4.4, with detailed results reported in Appendix B. Among all MTFs, inpatient utilization would have to increase (or decrease) by at least 28.2 percent on average for a change to be deemed systematic in nature rather than random. The average changes required for statistical significance were 15.3 percent, 10.4 percent, and 16.9 percent for outpatient utilization, drug

Table 4.4
Percentage Change in Outcome Required for Statistical Significance, Mean by MTF Size Quintile

Size Quintile	Inpatient Utilization	Outpatient Utilization	Drug Utilization	Total Cost
All	28.2	15.3	10.4	16.9
1 (smallest)	41.8	18.6	12.1	20.1
2	31.3	18.0	9.9	19.6
3	24.3	13.0	9.9	14.2
4	22.7	14.8	10.9	15.1
5 (largest)	20.7	11.8	9.2	15.4

NOTES: The confidence level is 95 percent. Size quintiles are defined by mean non-active duty Prime enrollment in FYs 2004–2005; see Table 4.3.

utilization, and total cost, respectively. Larger changes are required for inpatient utilization because its random variability is relatively great.[4]

For each outcome, the required changes tend to decrease in magnitude with MTF size. For example, outpatient utilization would have to increase (or decrease) by more than 18.6 percent on average for the smallest quintile of MTFs, but by only 11.8 percent for the largest quintile. This pattern is consistent with the earlier finding that the standard deviations of outcomes tended to decrease with MTF size, while their mean levels did not.

The reduction in required changes with size is especially pronounced for inpatient utilization, decreasing from 41.8 percent for the smallest MTFs to 20.7 percent for the largest. This pattern is consistent with the substantially larger changes required for inpatient utilization among all MTFs. These findings suggest that the enhanced opportunity for enrollees' inpatient care to balance out at larger MTFs may be especially important for inpatient utilization.

We are able to identify statistically significant differences in MTF outcomes from their 2004–2005 means during the quarters of FY 2006. Table 4.5 reports the frequency of MTF-quarters with statistically significant changes.[5] For example, among all MTFs, inpatient utilization was significantly different from the 2004–2005 mean in 10.1 percent of MTF-quarters; the frequencies were 26.1 percent, 19.3 percent, and 32.9 percent for outpatient utilization, drug utilization, and total cost, respectively. Given a 95 percent confidence level, we would expect 5 percent of observed changes to be deemed significant by chance when there is no change in the systematic determinants

---

[4]   Indeed, the magnitude of these changes may raise a concern that some of this variability is systematic rather than random. As we noted in Chapter Three, significant changes would correspond to changes in performance if performance were independent of the systematic factor that is contributing to the outcome's variability.

[5]   Statistical significance was determined based on two-sided hypothesis tests (Amemiya, 1994). Such tests are appropriate for identifying changes in outcomes in either direction, consistent with the project sponsor's interest in monitoring *any* changes in performance. Changes in only one direction can also be relevant, for example, if health care providers are rewarded for improvements in performance. In such a context, a one-sided test would be appropriate.

Table 4.5
Frequency of Statistically Significant Changes in MTF Outcomes During the Quarters of FY 2006 from Their Mean Levels in FYs 2004–2005, by MTF Size Quintile

Size Quintile	Inpatient Utilization (%)	Outpatient Utilization (%)	Drug Utilization (%)	Total Cost (%)
All	10.1	26.1	19.3	32.9
1 (smallest)	10.9	19.6	12.0	20.7
2	12.0	30.4	27.2	35.9
3	14.1	29.3	23.9	33.7
4	9.8	20.7	17.4	32.6
5 (largest)	3.4	30.7	15.9	42.0

NOTES: The confidence level is 95 percent. Size quintiles are defined by mean non-active duty Prime enrollment in FYs 2004–2005; see Table 4.3.

of outcomes, and 10.1 percent are found to be significant. As many as half (5%/10.1%) of the significant changes in inpatient utilization could therefore be "false positives." There are fewer false positives for the other outcomes; for example, the rate for total cost is no more than 15 percent (5%/32.9%). Unfortunately, we are unable to determine the false-negative rate.

The relatively low frequency of significant changes for inpatient utilization may be due in part to its greater variability. With inpatient utilization so variable, the required changes are large, and we may lack the power to discern systematic changes, resulting in many false negatives.

Another possible explanation for this low frequency is that the other outcomes tended to change more than inpatient utilization did in FY 2006, making these changes easier to discern. Indeed, in the MHS as a whole (not shown in the table), PMPM outpatient utilization, drug utilization, and total cost among non-active duty Prime enrollees grew by 8.2 percent, 4.7 percent, and 11.2 percent, respectively, while inpatient utilization grew by a more modest 3.7 percent. If we assess changes in MTF outcomes net of these MHS-wide trends, the frequency of significant changes in inpatient utilization is qualitatively unchanged, while the frequencies for the other outcomes decrease to levels comparable to inpatient utilization. This analysis is described in greater detail later in this chapter. The results suggest that performance assessment might have been as effective for inpatient utilization as for other outcomes, if MHS-wide trends had been comparable across outcomes. Yet these trends, and the differences among them, could have been the result of deteriorating performance across MTFs, as well as factors beyond the control of MTF leaders.

Table 4.5 also shows the frequency of significant changes by MTF size. Insofar as variability, and thus required changes, tends to decrease with size (Table 4.4), the frequency of significant changes may be lower at smaller MTFs. We find such a pattern for outpatient utilization, drug utilization, and total cost. For example, changes in total cost were significant in 20.7 percent of cases for the smallest MTFs but in 42.0 percent of cases for the largest. As a result, the share of significant changes that are false positives cannot be higher for the smallest MTFs than for the largest ones. For total cost, this share is as much as one-quarter (5%/20.7% = 24.2%) for the smallest MTFs, versus less than one-eighth (5%/42.0% = 11.9%) for the largest ones.

For inpatient utilization, by contrast, the frequency of significant changes increases from 10.9 percent among the smallest MTFs to 14.1 percent among moderately sized MTFs (quintile 3) but then decreases to only 3.4 percent at the largest MTFs. Differential trends in utilization growth by MTF size cannot account for this unexpected pattern. Utilization grew less at the smallest MTFs than at the largest ones (2.5 percent on average, versus 3.6 percent). One possible explanation is that the random variability of inpatient utilization changed between FYs 2004–2005 and FY 2006. Among the smallest MTFs, the estimated standard deviation was 61 percent larger on average in FY 2006 than in FYs 2004–2005; among the largest MTFs, the standard deviation increased by only 27 percent.[6] As a result, we are able to

---

[6]   These estimates are biased in finite samples. However, even with only the four quarters of FY 2006, the bias is modest. The statistics just presented should tend to understate the growth rate of the standard deviations, because the downward bias is greater in the period with fewer observations. In any event, this bias does not affect the identification of systematic changes. See Greene, 2003, p. 165.

discern fewer significant changes at the largest MTFs than we would have otherwise expected.[7]

Finally, we can describe the frequency with which there were significant differences from 2004–2005 levels during more than one quarter in 2006. Table 4.6 describes these frequencies. For inpatient utilization, there were significant changes at 33 MTFs. Among these MTFs, there were significant changes in all four quarters in 1 out of 33, or 3.0 percent, of cases. There were significant changes in costs at more MTFs, and they were more likely to be changes in multiple quarters. Of 62 MTFs with at least one significant change in costs, 10 of them, or 16.1 percent, had changes in all quarters.

Table 4.6
Number of MTFs with Statistically Significant Changes in Outcomes, by Number of Quarters in FY 2006 with Significant Changes

Number of Quarters with Significant Changes	Inpatient Utilization	Outpatient Utilization	Drug Utilization	Total Cost
0	81	54	62	52
1	23	22	25	14
2	8	18	19	18
3	1	19	7	20
4	1	1	1	10

NOTE: The confidence level is 95 percent.

---

[7] Appendix Table B.6 compares the variability of all outcomes in FYs 2004–2005 and FY 2006. Standard deviations increased over time for all outcomes. Only in the case of inpatient utilization did standard deviations grow by less at the largest MTFs than at the smallest ones. For outpatient utilization and total cost, they grew more at the largest MTFs than at the smallest ones, reinforcing the tendency (given their relatively low variability at large MTFs in FYs 2005–2006) to find more significant changes in these outcomes at large MTFs than at small ones.

Additional Issues

We now describe the results of some additional analyses of MTF PMPM utilization and costs among non-active duty TRICARE Prime enrollees. In particular, we analyze (1) statistically significant changes in these MTF outcomes based on a 90 percent confidence level; (2) changes in outcomes at the service level; (3) changes in MTF outcomes at an annual frequency; and (4) changes in MTF outcomes net of trends within the MHS, again averaged at a quarterly frequency.[8]

Lowering the Confidence Level

Our benchmark analysis of changes in MTF utilization and PMPM costs required 95 percent confidence in identifying statistically significant changes. We also consider a 90 percent confidence level. Figure 4.10 compares the average changes required for statistical significance under the lower and higher confidence levels. Figure 4.11 compares the frequency of significant changes in MTF outcomes during the quarters of FY 2006 from their mean levels in FYs 2004–2005.

Requiring less confidence decreases the changes required for statistical significance and increases the frequency of significant changes. That is, the false-negative rate declines (see Chapter Three). The cost, however, is that the false-positive rate rises. For example, whereas half (5%/10.1%) of the significant changes in inpatient utilization could have been false positives given 95 percent confidence, roughly two-thirds (10%/14.7%) of the changes could have been false positives given 90 percent confidence.

Assessing the Performance of the Military Services

We also assess the performance of services by aggregating the outcomes of MTFs operated by each of the services. Figures 4.12–4.15 compare the changes required for each of the services as a whole with those required on average of their individual MTFs, both based on a 95 per-

---

[8] We note here that OASD(HA) updated its PMPM data set during the course of this study. In earlier research conducted for OASD(HA), we found that the results of our preceding analysis of changes in MTF outcomes were qualitatively very similar across the two data sets.

Figure 4.10
Percentage Change in Outcome Required for Statistical Significance,
Mean by Confidence Level

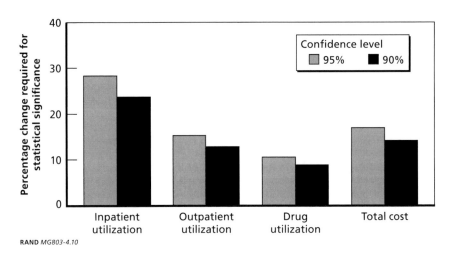

Figure 4.11
Frequency of Statistically Significant Changes in MTF Outcomes,
by Confidence Level

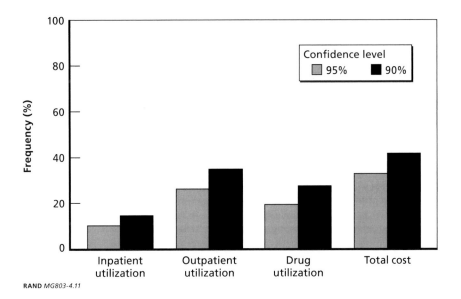

Figure 4.12
Percentage Change in Inpatient Utilization Required for Statistical Significance, Mean by Level of Analysis

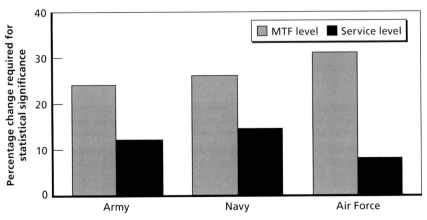

NOTE: The confidence level is 95%.
RAND *MG803-4.12*

Figure 4.13
Percentage Change in Outpatient Utilization Required for Statistical Significance, Mean by Level of Analysis

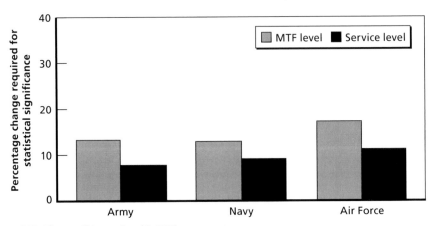

NOTE: The confidence level is 95%.
RAND *MG803-4.13*

Figure 4.14
Percentage Change in Drug Utilization Required for Statistical Significance,
Mean by Level of Analysis

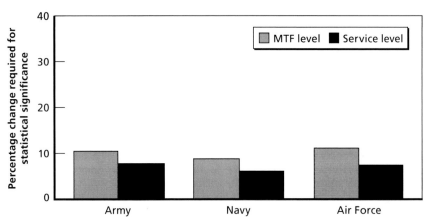

NOTE: The confidence level is 95%.

RAND *MG803-4.14*

Figure 4.15
Percentage Change in Total Cost Required for Statistical Significance,
Mean by Level of Analysis

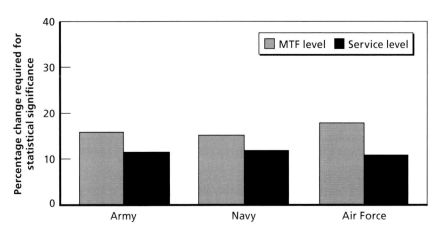

NOTE: The confidence level is 95%.

RAND *MG803-4.15*

cent confidence level. Smaller changes are required of the services as a whole, especially for inpatient care, because MTFs tend to balance out in their outcomes, just as enrollees at the same MTF do. These patterns suggest that performance assessment may be more effective at a service level. However, an exclusive focus on this level might deprive OASD(HA) of useful information concerning the performance of specific MTFs.

We cannot meaningfully analyze the frequency of significant changes during FY 2006, given the small number of services and quarters.

Assessing MTF Performance on an Annual Basis

We also assess MTF outcomes on an annual basis. An annual frequency of analysis may be more effective in identifying systematic changes in outcomes, because annual outcomes average over quarters within the year, potentially resulting in a decrease in their random variability.[9] Unfortunately, we cannot rely directly on estimates of the standard deviations of annual outcomes, because we have only two observations for each MTF's outcomes during FYs 2004–2005.

Appealing to the theoretical and empirical evidence in Chapters Three and Four as well as Appendix A concerning the impact of MTF size on the variability of outcomes, we specify and estimate several models of the relationships between the standard deviations of MTF outcomes, now measured at an annual frequency, and MTF size. In these regressions, the dependent variables are the estimated standard deviations of annual outcomes at MTFs.[10] For each outcome, the first specification includes only a constant, whose estimate corresponds to the mean value of the standard deviation. A second specification also includes the inverse square root of the number (in thousands) of non-active duty equivalent lives enrolled at MTFs, averaged over FYs 2004 and 2005. If the random components of enrollee-level outcomes

---

[9]   On the other hand, the small number of observations contributes to uncertainty about mean outcomes during 2004–2005, thereby requiring larger changes, all else equal.

[10]   Appendix A reports the formula used to estimate standard deviations based on observed outcomes.

have the same variability regardless of MTF size, then the variability of MTF-level outcomes decrease in proportion to this variable (see Appendix A). A third specification adds the number of enrollees, while a fourth further adds the squared number. These specifications flexibly relax the assumption of equal variability of enrollee outcomes across MTFs.[11] The results appear in Tables 4.7–4.10.[12] We prefer the specification whose adjusted R-squared is highest. In the annual analysis, we use the standard deviations predicted by the preferred specifications.

Figure 4.16 compares the required changes in annual outcomes with those in the benchmark analysis of outcomes measured at a quarterly frequency. Annual outcomes dampen the variability of quarterly outcomes, resulting in smaller required changes for statistical significance. This contrast is largest for inpatient utilization, whose average required change decreases from 28.2 percent to 14.8 percent.

The frequency of significant changes is compared in Figure 4.17. The frequency is substantially higher on an annual basis for all outcomes.[13] In relative terms, the increase is greatest for inpatient utilization, consistent with the relatively large reduction in required changes in Figure 4.12. As a result, the false-positive rate for inpatient utilization decreases from as much as half (5%/10.1%) to one-quarter (5%/20.2%); the false-negative rate is again unknown. The frequency of significant changes in costs may increase the least in relative terms because many changes were already identified on a quarterly basis. Altogether, these findings suggest that assessing MTF performance based on annual outcomes may be more effective. However, there is a potential tradeoff

---

[11] In Chapters Four and Five, we followed others in characterizing MTF size by the number of enrollees. In the annual analysis, the standard-deviation models account for the number of non-active duty equivalent lives, which is directly related to the outcome measures.

[12] We are unable to reject the hypothesis that the standard deviation is unrelated to MTF size in the case of prescription utilization. Because annual outcomes dampen the variability of quarterly outcomes, we would expect the size-variability relationship to be muted in comparison with the relationships illustrated in Figures 4.2–4.5 for outcomes measured at a quarterly frequency.

[13] This evidence cannot be attributed to the use of a model of the standard deviation in the annual analysis. The results of the benchmark analysis were qualitatively similar when a similar model was applied to outcomes measured at a quarterly frequency.

Table 4.7
Regressions of Estimated Standard Deviation of Annual Inpatient Utilization

	Specification			
	1	**2**	**3**	**4**
Parameter Estimate (Standard Error)				
Constant	0.400*** (0.034)	−0.003 (0.021)	−0.258*** (0.089)	−0.381* (0.222)
Inverse square root of number of non-active duty equivalent lives enrolled (thousands)	—	0.413*** (0.093)	0.026*** (0.009)	0.046 (0.034)
Number of non-active duty equivalent lives enrolled	—	—	0.954*** (0.204)	1.142*** (0.372)
Number squared	—	—	—	0.076 (0.051)
Other Statistics				
R-squared	0.000	0.000	0.073	0.076
Adjusted R-squared	0.000	−0.009	0.056	0.051
N			114	
Preferred Specification				
	No	No	Yes	No

NOTES: Independent variables are FY 2004–2005 means.

* denotes statistical significance at the 10 percent level.

** denotes statistical significance at the 5 percent level.

*** denotes statistical significance at the 1 percent level.

Table 4.8
Regressions of Estimated Standard Deviation of Annual Outpatient Utilization

	Specification			
	1	2	3	4
Parameter Estimate (Standard Error)				
Constant	2.883*** (0.234)	−0.413*** (0.140)	−1.002 (0.616)	−2.610* (1.532)
Inverse square root of number of non-active duty equivalent lives enrolled (thousands)	—	4.596*** (0.622)	0.060 (0.061)	0.322 (0.236)
Number of non-active duty equivalent lives enrolled	—	—	5.843*** (1.416)	8.306*** (2.572)
Number squared	—	—	—	0.091 (0.066)
Other Statistics				
R-squared	0.000	0.072	0.080	0.091
Adjusted R-squared	0.000	0.064	0.064	0.066
N		114		
Preferred Specification				
	No	Yes	No	No

NOTES: Independent variables are FY 2004–2005 means.
* denotes statistical significance at the 10 percent level.
** denotes statistical significance at the 5 percent level.
*** denotes statistical significance at the 1 percent level.

Table 4.9
Regressions of Estimated Standard Deviation of Annual Drug Utilization

	Specification			
	1	**2**	**3**	**4**
	Parameter Estimate (Standard Error)			
Constant	3.006*** (0.217)	0.128 (0.134)	0.178 (0.592)	0.502 (1.481)
Inverse square root of number of non-active duty equivalent lives enrolled (thousands)	—	2.475*** (0.596)	−0.005 (0.059)	−0.058 (0.228)
Number of non-active duty equivalent lives enrolled	—	—	2.369* (1.362)	1.872 (2.487)
Number squared	—	—	—	0.009 (−0.018)
	Other Statistics			
R-squared	0.000	0.008	0.008	0.009
Adjusted R-squared	0.000	−0.001	−0.010	−0.018
N		114		
	Preferred Specification			
	Yes	No	No	No

NOTES: Independent variables are FY 2004–2005 means.
* denotes statistical significance at the 10 percent level.
** denotes statistical significance at the 5 percent level.
*** denotes statistical significance at the 1 percent level.

Table 4.10
Regressions of Estimated Standard Deviation of Annual Costs

	Specification			
	1	**2**	**3**	**4**
Parameter Estimate (Standard Error)				
Constant	131.198***	2.429	−40.310*	−72.680
	(8.623)	(5.343)	(23.273)	(58.113)
Inverse square root of number of non-active duty equivalent lives enrolled (thousands)	—	121.134***	4.369*	9.630
		(23.769)	(2.317)	(8.956)
Number of non-active duty equivalent lives enrolled	—	—	211.749***	261.311***
			(53.495)	(97.559)
Number squared	—	—	—	0.036
				(0.010)
Other Statistics				
R-squared	0.000	0.002	0.033	0.036
Adjusted R-squared	0.000	−0.007	0.015	0.010
N			114	
Preferred Specification				
	No	No	Yes	No

NOTES: Independent variables are FY 2004–2005 means.

* denotes statistical significance at the 10 percent level.

** denotes statistical significance at the 5 percent level.

*** denotes statistical significance at the 1 percent level.

Figure 4.16
Percentage Change in Outcome Required for Statistical Significance,
Mean by Frequency of Analysis

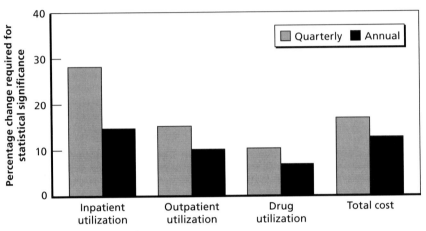

NOTE: The confidence level is 95%.

RAND *MG803-4.16*

between greater confidence in the results and more frequent assessment of outcomes.

Assessing MTF Performance Net of MHS Trends

We found in Chapter Four that the frequency of statistically significant changes was lower for inpatient utilization than for the other outcomes. One explanation is that we have relatively little power to discern systematic changes in patient utilization, because its random variability is large and thus so are the required changes.

Another explanation is that the other outcomes tended to change more between 2006 and 2004–2005 than inpatient utilization did, making differences easier to identify. Indeed, in the MHS as a whole, PMPM outpatient utilization, drug utilization, and total cost among non-active duty Prime enrollees grew by 8.2 percent, 4.7 percent, and 11.2 percent, respectively, while inpatient utilization grew by a comparatively modest 3.7 percent. This system-wide growth could be driven by a variety of factors common to MTFs. Benefit expansion and deteriorating performance are just two possibilities.

Figure 4.17
Frequency of Statistically Significant Changes in MTF Outcomes,
by Frequency of Analysis

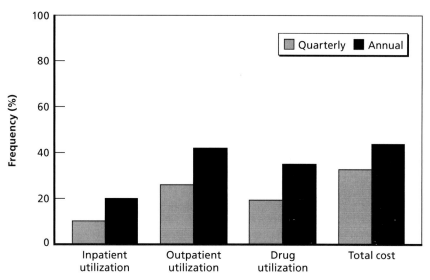

NOTE: The confidence level is 95%.
RAND *MG803-4.17*

We analyze changes in MTF outcomes net of MHS-wide trends. To do so, we adjust the mean levels of outcomes in FYs 2004–2005 for the growth rates just described. Figure 4.18 compares the frequencies of statistically significant changes in this analysis to the benchmark analysis in Chapter Four. The frequencies of changes in outpatient utilization, drug utilization, and total cost decrease substantially. This finding is not surprising, because we are now determining whether MTFs' changes in outcomes that grew throughout the MHS could have differed from these trends by chance. The frequency of significant changes in inpatient utilization is now quite similar to those for the other outcomes. This pattern suggests that performance assessment of inpatient utilization would have been as effective as the assessment of other outcomes, if MHS-wide trends were similar across outcomes.

As a matter of policy, assessing changes net of trend would not hold MTFs accountable for common factors. Insofar as these factors are beyond the control of MTFs, doing so is arguably "fair." Insofar as

Figure 4.18
Frequency of Statistically Significant Changes in MTF Outcomes, Observed
Changes Net of Military Health System Trend Versus Benchmark Analysis

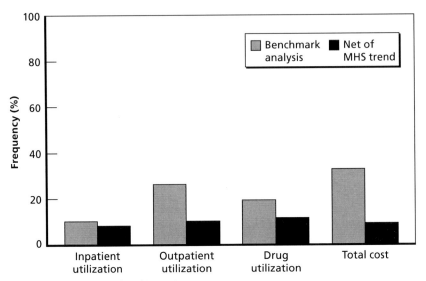

NOTE: The confidence level is 95%.

RAND *MG803-4.18*

these factors can be influenced, MTFs would not be held accountable for potentially important aspects of their performance. A similar observation would apply to assessments of each MTF in comparison with its "peer group," however defined.

In the next chapter, we use additional data to consider the role of catastrophic cases, such as organ transplants and low-birthweight deliveries. Excluding such cases might reduce the variability of outcomes, resulting in greater power to discern systematic changes.

CHAPTER FIVE

Performance Assessment and Catastrophic Cases

Catastrophic cases may contribute substantially to the random component of average PMPM utilization and costs among MTFs' enrollees. For example, organ transplants and low-birthweight deliveries may be resource-intensive enough that other cases cannot balance them out. If so, an observer could identify fewer statistically significant (that is, systematic) changes in MTFs' PMPM outcomes over time. This problem may be more acute at smaller MTFs, because there are relatively few enrollees and hence cases.

Excluding catastrophic cases from PMPM outcomes may therefore improve their effectiveness as performance measures. Under such an approach, performance assessment would be focused on non-catastrophic care. The practicality of excluding catastrophic cases using MHS data systems has been unclear.

In this chapter, we explore the practicality and usefulness of excluding catastrophic cases from PMPM outcomes. We focus on hospital care, because in Chapter Four we found inpatient utilization to be especially variable, and hospital care often plays an important role in catastrophic cases. We describe data sets relevant to catastrophic cases, explain our approach to identifying catastrophic inpatient care, and explore the impact of catastrophic care on the assessment of MTF performance with respect to inpatient utilization.

Relevant Data Sets

Data on inpatient cases are needed. The PMPM data set used in Chapter Four reports average inpatient utilization at MTFs. These measures aggregate across enrollees receiving different kinds of care, making it impossible to distinguish catastrophic from noncatastrophic utilization.

Appropriate disaggregate data are systematically captured by MHS data systems. In particular, the Standard Inpatient Data Record (SIDR) and Health Care Service Record–Institutional (HCSR-I) data sets include a wealth of information on each hospital admission in the direct and purchased-care systems, respectively; we describe the information used below. We have been able to obtain SIDR and HCSR-I data for FY 2004.

We also use Defense Enrollment Eligiblity Reporting System data sets to identify non-active duty Prime enrollees during this period. We supplement these data with data sets on MTF catchment areas in order to identify enrollment MTFs. In particular, we map each enrollee to a parent MTF based on the enrollee's zip code and sponsor service.[1]

Characterizing Catastrophic Inpatient Care

The SIDR and HCSR-I data sets report the diagnosis for each hospital patient's admission, as well as the associated Diagnosis-Related Group (DRG). International Classification of Disease codes characterize diagnoses, while DRGs reflect a broad diagnosis and, in some instances, a treatment regimen and/or outcome.

We use this information to identify catastrophic admissions. Catastrophic cases have been defined by International Classification of Disease codes.[2] Yet there does not appear to be any consensus definition; indeed, this practice is not even common. For each DRG, there

---

[1]   In the infrequent event that an enrollee's sponsor service was not reported and that the parent MTF within a zip code differed by service, we randomly assign the enrollee to one of the service's parent MTFs.

[2]   For example, see Tapei Branch of Bureau of National Health Insurance, 2008.

is a baseline Relative Weighted Product (RWP) that measures the typical resource use of such an admission; these values are reported in the SIDR and HCSR-I data sets. Our measure of inpatient utilization is also based on RWPs. Defining catastrophic admissions based on DRGs was therefore practical and well suited to the context.

We define admissions as catastrophic if their DRGs are among the highest with respect to RWPs. We consider as thresholds the top 1, 5, and 10 percent of DRGs in terms of resource use. Appendix D identifies the DRGs associated with catastrophic admissions. Examples include DRG 481 (bone marrow transplant) and DRG 602 (neonate, birthweight < 750 grams, discharged alive).

Under this definition, catastrophic admissions were rare as well as resource-intensive. Table 5.1 shows that, based on a threshold of the top 5 percent of DRGs by resource use, catastrophic admissions accounted for 0.7 percent of hospital admissions but 9.2 percent of total inpatient utilization. These admissions do not constitute a substantially larger share of utilization or admissions for enrollees at smaller MTFs, suggesting that MTFs of different sizes enroll beneficiaries who are similar in their catastrophic utilization. Even so, catastrophic admissions may still lead the random component of inpatient utilization to be large at MTFs with relatively few enrollees and admissions.

Table 5.1
Catastrophic Hospital Admissions as Share of Total Inpatient Utilization and Admissions Among Non-Active Duty Prime Enrollees, by Threshold for Catastrophic DRGs and MTF Size Quintile

Threshold/ Size Quintile	Percentage of Total Inpatient Utilization			Percentage of Total Hospital Admissions		
	Top 1%	Top 5%	Top 10%	Top 1%	Top 5%	Top 10%
All	3.4	9.2	16.9	0.1	0.7	2.3
1 (smallest)	3.1	9.6	17.1	0.1	0.7	2.2
2	3.1	9.9	17.6	0.1	0.8	2.3
3	3.1	8.4	16.2	0.1	0.7	2.3
4	3.8	9.2	16.8	0.2	0.8	2.4
5 (largest)	3.6	8.9	16.5	0.2	0.7	2.2

Assessing Changes in Inpatient Utilization with Catastrophic Admissions Excluded

Ideally, we could directly exclude catastrophic admissions from inpatient utilization, both in computing required changes based on FY 2004–2005 means and in identifying significant changes in FY 2006. The disaggregate data necessary were available only for FY 2004, however. We therefore explore the impact of excluding catastrophic admissions on the assessment of MTF performance with respect to inpatient utilization.

First, we regress total inpatient utilization during the four quarters on catastrophic utilization, as well as indicator variables for each MTF and the summer, fall, and winter quarters. Total and catastrophic inpatient utilization are found by summing the baseline RWPs associated with hospital admissions in SIDR and HCSR-I; we use the equivalent lives reported in the PMPM data set to compute average utilization, because equivalent lives are not directly reported in SIDR and HCSR-I.[3] The results of these regressions appear in Appendix E. We use these results to compute the magnitude of changes required for statistical significance at each MTF, based on the unexplained variability of total RWPs per equivalent life.[4]

As Figure 5.1 shows, the magnitudes of the required changes would decrease if catastrophic admissions were excluded. For a 95 percent

---

[3]   The measure of inpatient utilization in the current analysis differs from the measure used in the last chapter. In the current analysis, catastrophic admissions are defined according to the baseline RWPs associated with DRGs, and utilization therefore includes baseline RWPs. Utilization in Chapter Four was derived from the sponsor's PMPM data, which also includes outlier RWPs in inpatient utilization (Atkinson, 2007a). The correlation coefficient between the measures is 0.321. Our mapping of enrollee RWPs from the SIDR data to parent MTFs based on the Defense Enrollment Eligibility Reporting System (DEERS) introduced discrepancies between the numerators of the measures. We are able to compare the numbers of non-active duty enrollees under the linkage developed for the current analysis with the numbers reported in the PMPM data. The correlation coefficient is 0.908. We use equivalent lives from the PMPM data in the denominator of the current measure. Differences in the enrollee health status/risk between the numerator of the current measure and the PMPM data would exacerbate the discrepancies between the inpatient-utilization measures.

[4]   Because we are interested in the variability of the regression forecast, we account for the sampling variability of the regression parameters. See Greene, 2003.

Figure 5.1
Percentage Change in Noncatastrophic Inpatient Utilization Required
for Statistical Significance with 95 Percent Confidence, Mean by DRGs
Excluded Based on RWPs

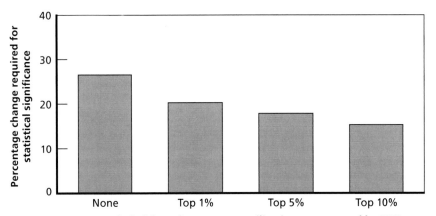

NOTE: The confidence level is 95%.

RAND *MG803-5.1*

confidence level, total inpatient utilization would have had to increase (or decrease) by nearly 26.5 percent on average (Table 4.4). By contrast, when we exclude admissions whose DRGs are in the top 5 percent by resource use, the required change in noncatastrophic care would be 17.9 percent at the average MTF. The required changes would be smaller still under the more inclusive 10 percent threshold.

Figure 5.2 shows the role played by MTF size, again based on quintiles of mean non-active duty enrollment during FYs 2004–2005. The decrease in required changes tends to be larger for smaller MTFs, especially the smallest. This pattern is not surprising. In Chapter Three, we explained that the impact of catastrophic cases on the variability of average MTF outcomes is likely to be larger at smaller MTFs, because there are relatively few enrollees whose health care needs can "balance out" at a point in time.

Next, we simulate the identification of significant changes in non-catastrophic inpatient utilization based on a thought experiment. In this experiment, we estimate the changes between 2004–2005 and

Figure 5.2
**Percentage Change in Noncatastrophic Inpatient Utilization Required for
Statistical Significance, Mean by DRGs Excluded and MTF Size Quintile**

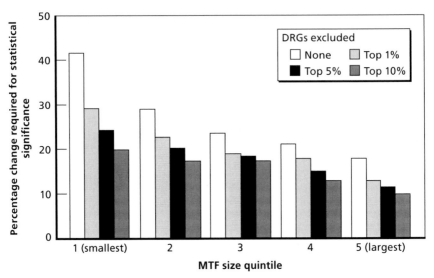

NOTE: The confidence level is 95%.
RAND *MG803-5.2*

2006 that would have been required for statistical significance. To do
so, we scale the change in total inpatient utilization required for each
MTF in Chapter Four by the average ratio of noncatastrophic changes
required to total changes required among MTFs in the same size quintile. We then assume that percentage changes in noncatastrophic utilization in 2006 equaled actual growth in total utilization. Finally, we
determine whether these estimated changes exceeded the estimated
requirements for statistical significance.

The results are illustrated in Figure 5.3. The frequency of significant changes increases substantially in general and for the smallest as
well as the largest MTFs in particular. Among all MTFs, the frequency
increased from 10.1 percent to 17.8 percent (not shown in Figure 5.3).
These findings, while tentative, suggest that assessing MTF performance with respect to noncatastrophic inpatient utilization could be
considerably more effective than assessing total utilization. It is pos-

Figure 5.3
Frequency of Statistically Significant Changes in Noncatastrophic Inpatient Utilization, by DRGs Excluded and MTF Size Quintile

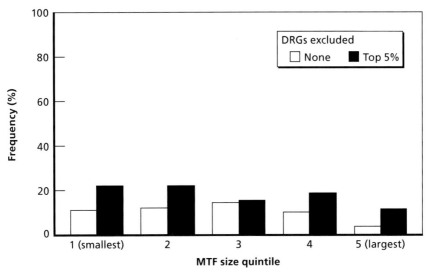

NOTE: The confidence level is 95%.
RAND *MG803-5.3*

sible, however, that the performance of MTF leaders in cost-effectively managing catastrophic care is critical but hard to assess.

CHAPTER SIX

Conclusions

The 1996 Quadrennial Defense Review charted a "Roadmap for Medical Transformation" that called for transforming business practices in the MHS through performance-based planning and financing. OASD(HA) has considered assessing the performance of the leaders of MTFs in managing care in a cost-effective manner. In particular, there has been potential interest in identifying changes in MTF utilization over time. Indeed, much like other health care payers, OASD(HA) has been monitoring PMPM utilization as well as costs among each MTF's Prime enrollees. Assessing performance based on average MTF outcomes might effectively align accountability and authority within MTFs and thereby help to sustain the benefit.

This monograph has described the results of this study of performance assessment of MTFs. In particular, we summarized the context for performance assessment in the MHS. We then offered an overview of performance assessment in nonmilitary health care, one that considered the potential usefulness and limitations of outcome-based measures of performance. This perspective on MTFs' PMPM outcomes motivated our quantitative analyses. We first determined the magnitude of statistically significant changes in PMPM utilization and costs. We then identified significant changes in these MTF outcomes. In doing so, we explored the roles of MTF size and catastrophic cases, which plausibly influence the random variability of outcomes and thus the effectiveness of performance assessment. We now present our conclusions.

Utilization and Costs Reflect MTF Performance and Other Factors

Our qualitative review of performance assessment in health care indicated that important outcomes are linked to performance in managing care. In our context, MTF leaders may systematically influence utilization and costs by providing more or less care, as well as by delivering care more or less efficiently. Changes in outcomes over time are, however, imperfect measures of performance.

Average MTF utilization and costs have a potentially important random component, because the need for and use of health care by an MTF's enrollees is somewhat random during any period. An observer must therefore decide whether changes are due to chance or to systematic factors, such as performance. Total certainty is impossible, and an observer will sometimes mistakenly conclude that a change is systematic in nature when in fact it is random ("false positives"). On other occasions, he will mistakenly conclude that a systematic change is random in nature ("false negatives"). An observer's confidence that a change is systematic diminishes with the outcome's random variability. The variability of average utilization and costs may be high at small MTFs, undermining the effectiveness of performance assessment. At an MTF with relatively few enrollees, the opportunity for their random health needs to balance out is limited. Outcomes may also be quite variable because of catastrophic cases, such as organ transplants. These cases may be too resource-intensive to be balanced by other cases.

Systematic factors other than performance further affect outcomes and thus their usefulness as performance measures. Potentially important examples include the structure of the TRICARE benefit, deployments of medical and other personnel, and the health status and typical health care needs of MTF enrollees. When such factors are stable over time at MTFs, systematic changes in outcomes must be due to changed performance. Yet these other factors may change over time. For example, there has been substantial turnover in the population of Prime enrollees at MTFs, potentially resulting in changes in the overall health status/risk of MTF enrollees. While PMPM outcomes account for health risk based on age, gender, and beneficiary category, such risk adjustment is imperfect. Changes in outcomes would be unreliable measures of performance if outcomes did not account for other systematic factors that are not independent of performance. Even if other factors were independent, they contribute to the variability of outcomes, potentially undermining confidence in assessments of performance. In any event, the study scope did not include a quantitative analysis of systematic bias in MTF performance assessment.

MTF Utilization and Costs Are Variable, But Systematic Changes Can Be Identified

We used OASD(HA)'s PMPM data set to assess MTF utilization and costs from FY 2004 through FY 2006. We analyzed average inpatient utilization, outpatient utilization, drug utilization, and total cost among each MTF's Prime enrollees, with active-duty personnel excluded because of deployment-related data concerns.

We found that these MTF outcomes varied substantially during 2004 and 2005, as measured by their standard deviations. We therefore investigated whether systematic changes from 2004–2005 could be identified in 2006. For each outcome at each MTF, we determined the magnitude of the change required for 95 percent confidence that the change is statistically significant, that is, systematic in nature. The required changes ranged from 10.4 percent on average for drug utilization to 28.2 percent for inpatient utilization. Large changes were required for inpatient utilization because this outcome was especially variable.

The frequency of significant changes in 2006 ranged from 10.1 percent for inpatient utilization to 32.9 percent for costs. The low frequency for inpatient utilization is not attributable to large required changes but rather to low growth, relative to the other outcomes, throughout the MHS. Given a 95 percent confidence level, 5 percent of changes would happen to be significant if systematic factors were unchanged. As many as half of the significant changes in inpatient utilization could therefore be false positives. For total cost, the false-positive rate would be no greater than 15 percent. Unfortunately, we cannot quantify false-negative rates. If the confidence level were lower, more changes would have been significant, resulting in more false positives but fewer (though still some) false negatives. Whether these false-positive rates are unacceptably high is a policy judgment.

Outcomes Are More Variable at Smaller MTFs, Yet MTF Size Was Not Consistently Related to the Frequency of Systematic Changes

We investigated the impact of MTF size on the effectiveness of performance assessment. We defined MTF size by the average number of non-active duty enrollees during FYs 2004–2005 and created five groupings of similarly sized MTFs.

As expected, outcomes were substantially more variable at smaller MTFs. Required changes were larger as a result. For inpatient utilization, required changes averaged 41.8 percent at the smallest MTFs, versus 20.7 percent at the largest. There was a similar, albeit less extreme, pattern for the other outcomes.

The frequency of systematic (that is, statistically significant) changes also varied with MTF size. The frequency of changes in total cost, for example, increased from 20.7 percent on average for the smallest MTFs to 42.0 percent for the largest. Outpatient and drug utilization exhibited similar, though less pronounced, patterns. The frequency for inpatient utilization first increased from 10.9 percent at the smallest MTFs to 14.1 percent at moderately sized MTFs, then decreased to 3.4 percent among the largest. Rapid growth at the small MTFs cannot explain this finding, as inpatient utilization grew faster at the largest MTFs than at the smallest. Overall, performance assessment is not consistently less effective at small MTFs.

Excluding Catastrophic Inpatient Care Is Practical and Potentially Useful

Finally, we excluded catastrophic cases from inpatient utilization for several reasons. Inpatient care is often important in treating these cases. Inpatient utilization was especially variable. Necessary data on hospital admissions in the MHS were also readily available for 2004.

We characterized a hospital admission as catastrophic if the DRG associated with the admission was relatively resource-intensive. We considered the top 1 percent, top 5 percent, and top 10 percent of diagnoses as thresholds for catastrophic admissions.

Catastrophic accounted for a disproportionate share of resources: 0.7 percent of admissions but 9.2 percent of inpatient utilization, based on the 5 percent threshold. This pattern varied little by MTF size.

Based on this threshold, the required change in noncatastrophic utilization was 17.9 percent on average, versus 28.2 percent for total inpatient utilization. The required change would be even smaller with the more inclusive 10 percent threshold.

With one year of data, we could not directly identify statistically significant changes in noncatastrophic utilization. We were able to simulate this frequency under some simplifying assumptions. Based on the 5 percent threshold for catastrophic admissions, the frequency of significant changes was 17.8 percent, versus 10.1 percent for total inpatient utilization. It is possible that catastrophic care is a critical aspect of MTF performance that is, unfortunately, hard to assess.

Potential Refinements

Changes over time in average utilization and costs at MTFs may, despite their limitations, be useful measures of MTF performance. The practical importance of systematic bias in these measures would seem to merit investigation. If biases are substantial, there may be feasible modifications of the measures that would enhance their reliability. In addition, our review of performance assessment in health care suggests that alternative measures might complement average MTF outcomes.

Such alternative approaches as targeted assessments of episodes or clinicians could help to diagnose MTF performance problems more reliably and to treat them more effectively. On the other hand, implementing such metrics may not be practical in the MHS context, given current data systems. Much remains unknown about these issues.

Some Statistical Issues in Performance Assessment

This appendix supplements the discussion in Chapter Three of statistical issues relating to performance assessment. We now present a more formal framework for concluding whether changes in outcomes over time are systematic or random in nature. We also offer some numerical examples of the potential roles that MTF size and catastrophic cases play in drawing these conclusions.

For concreteness, we will consider inpatient utilization at an MTF. (The framework applies equally to other outcomes.) Suppose that enrollee e's utilization during period t is

$$u_{e,t} = -p_t + u + v_{e,t}, \tag{1}$$

in which p_t is the MTF's performance in cost-effectively managing care, u is an enrollee's typical need for and use of inpatient care based on his or her health status/risk, and $v_{e,t}$ is a random component of the enrollee's utilization with a mean of 0. Note that better (that is, higher) performance is defined as lower utilization, given the negative sign that precedes p_t.

In this appendix, we assume for simplicity that performance is the only systematic determinant of outcomes that varies over time.[1]

---

[1] The model also assumes that performance and the random component of utilization are additively separable from all other factors (including each other). This assumption is common in performance assessment (see, e.g., Goldstein and Spiegelhalter, 1996).

In reality, changes in the health status/risk of enrollees systematically affect utilization. The quantitative analyses in the body of the report incorporated health status/risk based on age, gender, and beneficiary category. In addition, Chapter Three offered a theoretical perspective on the possibility that unobserved systematic factors lead to changes in outcomes over time.

OASD(HA) has been monitoring PMPM utilization and cost (see Chapters Two and Three). Based on equation 1, average inpatient utilization among the MTF's enrollees is

$$\bar{u}_t \equiv$$

$$\frac{\sum_e u_{e,t}}{n} = \frac{\sum_e -p_t + u + v_{e,t}}{n} = -p_t + u + \frac{\sum_e v_{e,t}}{n} \quad (2)$$

$$\equiv -p_t + u + \bar{v}_t,$$

in which n denotes the number of enrollees at the MTF, which is assumed to be constant over time for simplicity. The outcomes analyzed in this monograph are averaged over equivalent lives, not enrollment (see Chapter Three); we ignore this distinction here, having already assumed that health status/risk is identical across enrollees and over time. Equation 2 shows that average utilization depends on MTF performance and the average value of enrollees' random utilization, $\bar{v}_t$.

Based on equation 2, the change in average inpatient utilization over time is

$$\Delta \bar{u}_t \equiv \bar{u}_t - \bar{u}_{t-1} = -(p_t - p_{t-1}) + (\bar{v}_t - \bar{v}_{t-1}) \equiv -\Delta p_t + \Delta \bar{v}_t. \quad (3)$$

Thus, a change in utilization could be caused by a change either in performance or in the random utilization of enrollees on average.

It is then possible to test the hypothesis that an observed change in inpatient utilization is random, rather than systematic, in nature. The hypothesis is that $\Delta p_t = 0$, and the likelihood of this hypothesis equals the likelihood of the event $\Delta \bar{v}_t = \Delta \bar{u}_t$. This latter likelihood is determined by the probability distribution of $\Delta \bar{v}_t$.

This distribution converges to a normal distribution as the number of enrollees approaches infinity, if $v_{e,t}$ is distributed independently and identically across enrollees and over time (Amemiya, 1994).[2] The mean of $\bar{v}_t$ under its asymptotic distribution is 0, and the variability is

$$SD(\Delta\bar{v}_t) = \sqrt{\frac{2}{n}} \cdot SD(v_{e,t}), \qquad (4)$$

where $SD(x)$ is the standard deviation of a random variable x. The standard deviation is defined as

$$\sqrt{E(x - Ex)^2}.$$

It can be estimated (as we do)[3] by

$$\sqrt{\frac{\sum_i (x_i - \bar{x}_i)^2}{(n-1)}},$$

with

$$\bar{x}_i = \frac{\sum_i x_i}{n}.$$

The assumption that $v_{e,t}$ is independently and identically distributed across enrollees and over time means that $SD(v_{e,t})$ and $SD(\Delta\bar{v}_t)$ are constants.

---

[2]   Such convergence can hold for the average of independent random variables even when their means and variances (as defined below) differ (Amemiya, 1994). In our benchmark analysis, we do not place any restrictions on the means or variances of average utilization, etc., across outcomes or MTFs, because we analyze each outcome at each MTF separately. With means or variances that vary over time during a period, convergence under more general conditions becomes necessary and may in fact hold. In this event, we are comparing average performance during the baseline period of FYs 2004–2005 with subsequent performance.

[3]   The (n – 1) term in the denominator ensures that the estimated variance is unbiased.

An observer can reject the hypothesis that an observed change $\Delta\bar{u}_t$ is due to chance with a degree of confidence determined by $SD(\Delta\bar{v}_t)$. The confidence level can be found by comparing the statistic

$$z = \frac{\Delta\bar{u}_i}{SD(\Delta\bar{v}_i)}$$

with a standard normal distribution table (see Amemiya, 1994).

Alternatively, an observer could specify the degree of confidence required and then determine whether an observed change exceeds the change required for this confidence level. For example, if 95 percent confidence were required, a change of $1.96 \cdot SD(\Delta\bar{v}_t)$—whether an increase or a decrease—would be required to reject the hypothesis that the change is due to chance. If, on the contrary, utilization during period t were within the confidence interval

$$\left[\bar{v}_{t-1} - 1.96 \cdot SD(\Delta\bar{v}_t), \bar{v}_{t-1} + 1.96 \cdot SD(\Delta\bar{v}_t)\right],$$

this hypothesis could not be rejected.[4]

The required change (as well as the confidence interval) shrinks with the number of enrollees at an MTF. Table A.1 considers MTFs with 1,000, 10,000, and 100,000 enrollees. Inpatient utilization (RWPs) is set equal to its monthly average of 0.00615 per equivalent life (6.15 per thousand lives, as reported in Table 4.1) during FYs 2004–2005, so that this example is not confounded by a relationship between utilization and MTF size. We assume for numerical concreteness that $SD(u_{e,t}) = 0.05$. Comparing MTFs with 1,000 and 10,000 enrollees, the required change is 0.00310 (or 50.4 percent) at the smaller MTF, versus 0.00098 (or 15.9 percent) at the larger one. Comparing MTFs with 10,000 and 100,000 enrollees, the required change is again smaller at the larger MTF, though by a lesser increment

$$(0.00098 - 0.00031 < 0.00310 - 0.00098).$$

---

[4]   Strictly speaking, the statement that a hypothesis of no change can be rejected if the observed change lies outside the confidence interval surrounding the original utilization level is valid under the normal distribution that obtains here, but not for other distributions.

The intuition in this example is that there is greater opportunity for individual enrollees' random needs for inpatient care to balance out when there are more enrollees at an MTF.

Table A.1 also shows that smaller changes are required when the required level of confidence is smaller.[5] That is, there will be fewer false negatives with 90 percent confidence. The cost, however, is that the false-positive rate will increase by a magnitude that varies with the context.

The required change may also shrink when catastrophic cases, such as organ transplants, are excluded from inpatient utilization. Because these cases occur somewhat randomly, an enrollee's random need for inpatient care during a period can be decomposed:

$$v_{e,t} = v_{e,t}^c + v_{e,t}^{nc}, \tag{5}$$

in which the superscript c denotes catastrophic inpatient care, while the superscript nc denotes noncatastrophic inpatient care. We then have

$$SD(v_{e,t}) = \sqrt{Var(v_{e,t}^c) + 2Cov(v_{e,t}^c, v_{e,t}^{nc}) + Var(v_{e,t}^{nc})},$$

in which $Var(x)$ is the variance of a random variable x, and $Cov(x,y)$ is the covariance of random variables x and y. The variance of x is defined to be equal to the square of its standard deviation, while the covariance of x and y is defined as $E\left[(x - Ex)(y - Ey)\right]$.

Catastrophic cases may contribute substantially to the standard deviation of total utilization. Even if catastophic cases are rare, utilization is quite high on those occasions.

Excluding catastrophic cases from inpatient utilization transforms the standard deviation of $\Delta \bar{u}_t$ by a factor of

$$\sqrt{\frac{Var(v_{e,t}^{nc})}{Var(v_{e,t}^c) + Var(v_{e,t}^{nc})}} < 1,$$

---

[5] For a 90 percent confidence level, a change of $1.645 \cdot SD(\Delta \bar{v}_t)$ is required for statistical significance.

Table A.1
Random Variability of Inpatient Utilization and Confidence Intervals for Statistically Significant Changes, by MTF Size and Confidence Level

Number of Enrollees (N)	Inpatient Utilization per Enrollee ($\bar{u}_t$)	Standard Deviation of Enrollee Utilization ($SD(u_{t,e})$)	Standard Deviation of Utilization per Enrollee ($SD(\bar{u}_t)$)	95 Percent Confidence			90 Percent Confidence		
				Absolute Change	Confidence Interval	Percentage Change	Absolute Change	Confidence Interval	Percentage Change
1,000	0.00615	0.05	0.00158	0.00310	[0.00305, 0.00925]	50.4	0.00260	[0.00355, 0.00875]	42.3
10,000	0.00615	0.05	0.00050	0.00098	[0.00517, 0.00713]	15.9	0.00082	[0.00533, 0.00697]	13.4
100,000	0.00615	0.05	0.00016	0.00031	[0.00584, 0.00646]	5.0	0.00026	[0.00589, 0.00641]	4.2

NOTES: Confidence interval and percentage changes are based on a level of utilization in the prior period equal to average inpatient utilization (RWPs) per 1,000 equivalent lives in FYs 2004–2005.

assuming that $Cov(x,y) = 0$. If, for example, the variance of catastrophic and noncatastrophic care were equal, then the required change, now in terms of noncatastrophic utilization, would shrink with the standard deviation by nearly 30 percent $[100 \cdot (1 - 2^{-1/2})]$. At the MTF with 10,000 enrollees in Table A.1, the required change, in percentage terms, would decrease from 15.9 percent to 11.2 percent. With required changes that are relatively large at smaller MTFs, excluding catastrophic cases would bring greater gains in statistical power, and thus the effectiveness of performance assessment, for these MTFs.

As an alternative to excluding catastrophic cases, all cases of inpatient utilization could be included but normalized (i.e., divided) by a measure of their typical resource-intensiveness. Catastrophic and noncatastophic cases would then have similar influence on the level of average inpatient utilization. Their impacts on the random variability of average utilization would turn on their variability around their typical resource-intensiveness. Such an approach could identify systematic changes with greater frequency than a PMPM metric, without ignoring a potentially important group of cases, as a measure that excludes catastrophic cases would. This alternative, while potentially useful, was beyond the scope of this study.

Detailed Results of Benchmark Analysis of MTF Outcomes

This appendix reports the detailed results of our benchmark analysis of MTF Prime enrollment and equivalent lives as well as PMPM outcomes, measured at a quarterly frequency and averaged over FYs 2004 and 2005. Chapter Four describes this analysis.

Table B.1 describes mean MTF size under the measures listed in Table 4.2, ordered by the number of non-active duty enrollees.

Tables B.2–B.5 report the mean of each outcome at each MTF as well as its standard deviation. The bounds describe the ends of the confidence intervals discussed in Appendix A, given a 95 percent confidence level; they therefore define the magnitudes of the changes required for a change to be statistically significant.

Finally, Table B.6 compares the estimated standard deviations of MTF outcomes in FYs 2004–2005 with their levels in FY 2006, based on MTF size. As in Table 4.3, we create five groups of similarly sized MTFs based on mean non-active duty enrollment.

Table B.1
TRICARE Prime Enrollment and Equivalent Lives, FYs 2004–2005

MTF Name	Parent DMIS ID	Enrollment		Equivalent	
		Total	Excluding Active Duty	Total	Excluding Active Duty
460th MED GRP-BUCKLEY AFB	7200	4,181	824	3,989	821
71st MED GRP-VANCE	338	3,837	2,496	4,046	2,932
47th MED GRP-LAUGHLIN	114	4,094	2,670	4,540	3,318
PATTERSON AHC-FT. MONMOUTH	81	12,547	3,513	11,875	3,832
14th MED GRP-COLUMBUS	74	5,138	3,528	5,736	4,379
97th MED GRP-ALTUS	97	5,917	4,353	6,136	4,747
17th MED GRP-GOODFELLOW	364	7,187	4,791	7,189	5,011
319th MED GRP-GRAND FORKS	93	7,703	4,818	7,334	4,879
354th MED GRP-EIELSON	203	8,120	4,981	7,291	4,572
NHC ANNAPOLIS	306	11,225	5,257	10,829	5,797
66th MED GRP-HANSCOM	310	8,312	5,296	8,291	5,331
61st MED SQUAD-LOS ANGELES	248	8,687	5,378	8,860	5,719
9th MED GRP-BEALE	15	9,059	5,563	8,830	5,787
579th MED GROUP-BOLLING	413	11,361	5,793	11,812	5,947
KELLER ACH-WEST POINT	86	12,061	5,816	11,633	6,216
30th MED GRP-VANDENBERG	18	9,287	6,113	9,287	6,551
27th MED GRP-CANNON	85	9,824	6,236	9,692	6,681
341st MED GRP-MALMSTROM	77	10,440	6,645	10,481	7,263
5th MED GRP-MINOT	94	11,671	6,703	10,862	6,638
90th MED GRP-F.E. WARREN	129	10,285	6,704	10,058	6,944
WEED ACH-FT. IRWIN	131	11,799	6,800	10,590	6,473
95th MED GRP-EDWARDS	19	10,262	6,980	9,912	7,135
23rd MED GRP-MOODY	50	11,348	7,187	10,795	7,237
49th MED GRP-HOLLOMAN	84	11,125	7,403	11,879	8,632
437th MED GRP-CHARLESTON	356	12,662	7,742	12,775	8,369

Table B.1—Continued

MTF Name	Parent DMIS ID	Enrollment		Equivalent Lives	
		Total	Excluding Active Duty	Total	Excluding Active Duty
366th MED GRP-MOUNTAIN HOME	53	12,335	7,959	12,437	8,657
43rd MEDICAL GROUP-POPE	335	13,432	8,007	12,762	8,059
NHC PATUXENT RIVER	68	11,478	8,022	11,323	8,343
509th MED GRP-WHITEMAN	76	11,937	8,134	11,761	8,485
28th MED GRP-ELLSWORTH	106	11,867	8,167	12,364	9,182
22nd MED GRP-MCCONNELL	59	11,748	8,415	12,058	9,101
BASSETT ACH-FT. WAINWRIGHT	5	16,125	8,441	14,397	8,356
15th MED GRP-HICKAM	287	13,661	8,588	13,198	8,260
1st SPEC OPS MED GRP-HURLBURT	7139	16,434	8,653	14,862	8,212
7th MED GRP-DYESS	112	14,561	9,197	14,298	9,816
62nd MED GRP-MCCHORD	395	13,835	9,421	14,774	10,894
WALTER REED AMC-WASHINGTON DC	37	23,720	9,813	28,020	12,458
305th MED GRP-MCGUIRE	326	16,942	9,896	16,109	9,481
314th MED GRP-LITTLE ROCK	13	16,260	10,081	16,373	10,956
4th MED GRP-SEYMOUR JOHNSON	90	14,755	10,090	14,981	10,890
92nd MED GRP-FAIRCHILD	128	13,789	10,158	14,821	11,653
FOX AHC-REDSTONE ARSENAL	1	11,475	10,170	13,722	12,426
NH TWENTYNINE PALMS	30	13,530	10,172	13,600	10,894
436th MED GRP-DOVER	36	15,414	10,487	16,021	11,587
NH BEAUFORT	104	13,475	10,490	13,999	11,418
20th MED GRP-SHAW	101	16,776	10,990	16,955	11,816
82nd MED GRP-SHEPPARD	113	15,199	11,030	15,719	11,972
LYSTER AHC-FT. RUCKER	3	15,096	11,500	16,060	12,893
2nd MED GRP-BARKSDALE	62	17,997	11,733	18,531	13,102
325th MED GRP-TYNDALL	43	16,152	11,743	17,451	13,524

Table B.1—Continued

MTF Name	Parent DMIS ID	Enrollment		Equivalent Lives	
		Total	Excluding Active Duty	Total	Excluding Active Duty
KENNER AHC-FT. LEE	122	15,598	11,913	16,051	12,401
R W BLISS AHC-FT. HUACHUCA	8	16,469	12,376	17,807	14,120
NH CHARLESTON	103	20,999	12,647	21,886	15,435
NH CORPUS CHRISTI	118	19,778	12,962	20,636	14,859
MUNSON AHC-FT. LEAVENWORTH	58	17,024	12,975	17,450	13,652
NH OAK HARBOR	127	16,035	13,050	16,326	13,788
NH LEMOORE	28	16,514	13,070	16,730	13,823
45th MED GRP-PATRICK	46	15,823	13,158	18,654	16,206
BAYNE-JONES ACH-FT. POLK	64	22,705	13,568	21,676	14,284
NH CHERRY POINT	92	15,756	13,651	16,666	14,903
75th MED GRP-HILL	119	19,697	13,675	19,383	14,156
NHC GREAT LAKES	56	19,715	13,829	19,409	13,916
42nd MEDICAL GROUP-MAXWELL	4	18,047	13,891	19,193	14,913
GUTHRIE AHC-FT. DRUM	330	31,409	14,207	27,357	13,353
78th MED GRP-ROBINS	51	21,865	15,528	21,899	16,096
377th MED GRP-KIRTLAND	83	20,294	15,587	22,809	18,516
L. WOOD ACH-FT. LEONARD WOOD	75	23,198	15,925	23,395	17,015
21st MED GRP-PETERSON	252	22,695	16,140	22,853	16,629
72nd MED GRP-TINKER	96	25,611	16,890	26,123	18,398
IRWIN ACH-FT. RILEY	57	30,477	17,569	28,375	18,047
MONCRIEF ACH-FT. JACKSON	105	22,563	17,968	24,408	19,693
12th MED GRP-RANDOLPH	366	23,025	18,370	26,230	21,583
NHC QUANTICO	385	27,253	18,787	25,279	18,348
355th MED GRP-DAVIS MONTHAN	10	26,363	18,931	28,395	21,745

Table B.1—Continued

MTF Name	Parent DMIS ID	Enrollment		Equivalent Lives	
		Total	Excluding Active Duty	Total	Excluding Active Duty
10th MED GROUP-USAF ACADEMY CO	33	25,441	19,295	28,326	22,838
IRELAND ACH-FT. KNOX	61	33,347	19,456	32,866	20,336
56th MED GRP-LUKE	9	27,580	20,687	30,867	24,742
55th MED GRP-OFFUTT	78	28,113	20,703	28,390	21,609
REYNOLDS ACH-FT. SILL	98	30,834	21,423	29,091	21,439
NAVAL HEALTH CARE NEW ENGLAND	6223	36,059	21,563	34,735	22,576
NHC HAWAII	280	30,211	21,801	28,908	21,301
1st MED GRP-LANGLEY	120	32,005	22,549	31,150	21,966
MCDONALD AHC-FT. EUSTIS	121	29,189	22,755	30,256	24,208
81st MED GRP-KEESLER	73	28,732	22,857	32,117	26,442
375th MED GRP-SCOTT	55	30,368	23,448	32,340	25,355
79th MED GRP-ANDREWS	66	31,453	24,128	33,294	25,584
3rd MED GRP-ELMENDORF	6	31,806	24,197	31,545	24,744
NNMC BETHESDA	67	41,233	24,604	46,120	29,321
6th MED GRP-MACDILL	45	32,747	24,921	37,943	30,303
96th MED GRP-EGLIN	42	34,199	25,121	35,874	27,923
88th MED GRP-WRIGHT-PATTERSON	95	34,487	27,571	39,518	32,744
WILLIAM BEAUMONT AMC-FT. BLISS	108	40,111	27,828	42,187	31,068
60th MED GRP-TRAVIS	14	37,298	27,911	44,189	35,357
99th MED GRP-O'CALLAGHAN HOSP	79	37,095	28,608	41,956	34,525
NH CAMP LEJEUNE	91	32,962	29,004	32,948	29,458
EVANS ACH-FT. CARSON	32	46,799	29,370	45,461	31,511
MARTIN ACH-FT. BENNING	48	43,937	29,844	43,720	31,962
NH BREMERTON	126	35,107	29,972	35,848	31,330

Table B.1—Continued

MTF Name	Parent DMIS ID	Enrollment		Equivalent Lives	
		Total	Excluding Active Duty	Total	Excluding Active Duty
KIMBROUGH AMB CAR CEN-FT. MEADE	69	44,664	31,462	45,131	31,983
TRIPLER AMC-FT. SHAFTER	52	50,965	32,020	48,183	32,029
BROOKE AMC-FT. SAM HOUSTON	109	38,504	32,999	46,808	40,938
WINN ACH-FT. STEWART	49	59,761	33,390	55,787	33,608
NH PENSACOLA	38	49,236	33,463	50,398	36,038
EISENHOWER AMC-FT. GORDON	47	52,880	35,266	57,598	40,293
59th MED WING-LACKLAND	117	51,235	39,350	58,900	46,961
BLANCHFIELD ACH-FT. CAMPBELL	60	65,845	39,433	60,294	39,191
NH JACKSONVILLE	39	59,685	44,730	58,494	45,305
NH CAMP PENDLETON	24	58,672	48,161	60,639	51,793
MADIGAN AMC-FT. LEWIS	125	84,234	52,561	86,293	59,210
DARNALL AMC-FT. HOOD	110	109,119	56,724	97,505	54,139
WOMACK AMC-FT. BRAGG	89	100,686	58,808	92,404	57,943
DEWITT ACH-FT. BELVOIR	123	78,853	67,780	87,535	76,162
NMC SAN DIEGO	29	105,480	81,414	110,673	89,339
NMC PORTSMOUTH	124	121,079	85,109	116,328	83,240

NOTE: These statistics are measured at a quarterly frequency and averaged over FYs 2004–2005.

Table B.2
Monthly Inpatient Utilization (RWPs) per 1,000 Equivalent Lives, FYs 2004–2005

MTF Name	Parent DMIS ID	Mean	Standard Deviation	95% Lower Bound	95% Upper Bound
460th MED GRP-BUCKLEY AFB	7200	5.47	1.40	2.56	8.38
71st MED GRP-VANCE	338	7.70	2.37	2.76	12.63
47th MED GRP-LAUGHLIN	114	5.47	2.68	–0.10	11.03
PATTERSON AHC-FT. MONMOUTH	81	5.84	0.57	4.66	7.02
14th MED GRP-COLUMBUS	74	5.49	0.96	3.48	7.49
97th MED GRP-ALTUS	97	6.63	0.94	4.68	8.57
17th MED GRP-GOODFELLOW	364	5.75	0.47	4.78	6.73
319th MED GRP-GRAND FORKS	93	6.72	2.14	2.28	11.16
354th MED GRP-EIELSON	203	5.50	1.29	2.82	8.17
NHC ANNAPOLIS	306	2.41	0.71	0.94	3.88
66th MED GRP-HANSCOM	310	5.38	0.98	3.35	7.41
61st MED SQUAD-LOS ANGELES	248	5.31	1.07	3.08	7.54
9th MED GRP-BEALE	15	5.82	1.28	3.15	8.49
579th MED GROUP-BOLLING	413	3.47	0.49	2.46	4.48
KELLER ACH-WEST POINT	86	6.41	0.67	5.02	7.81
30th MED GRP-VANDENBERG	18	5.04	0.77	3.44	6.64
27th MED GRP-CANNON	85	6.84	1.25	4.24	9.43
341st MED GRP-MALMSTROM	77	6.64	1.07	4.40	8.87
5th MED GRP-MINOT	94	6.07	1.15	3.67	8.46
90th MED GRP-F.E. WARREN	129	5.91	0.91	4.02	7.80
WEED ACH-FT. IRWIN	131	7.73	1.52	4.58	10.88
95th MED GRP-EDWARDS	19	5.39	0.68	3.98	6.81
23rd MED GRP-MOODY	50	4.89	1.07	2.68	7.11
49th MED GRP-HOLLOMAN	84	6.84	1.08	4.59	9.09
437th MED GRP-CHARLESTON	356	5.65	0.96	3.66	7.64

Table B.2—Continued

MTF Name	Parent DMIS ID	Mean	Standard Deviation	95% Lower Bound	95% Upper Bound
366th MED GRP-MOUNTAIN HOME	53	6.72	1.01	4.61	8.82
43rd MEDICAL GROUP-POPE	335	4.75	0.69	3.32	6.18
NHC PATUXENT RIVER	68	4.54	0.73	3.01	6.06
509th MED GRP-WHITEMAN	76	6.28	0.57	5.11	7.46
28th MED GRP-ELLSWORTH	106	5.93	1.22	3.40	8.46
22nd MED GRP-MCCONNELL	59	6.32	0.68	4.91	7.73
BASSETT ACH-FT. WAINWRIGHT	5	7.45	1.00	5.36	9.54
15th MED GRP-HICKAM	287	3.33	0.46	2.37	4.29
1st SPEC OPS MED GRP-HURLBURT	7139	3.84	0.49	2.83	4.86
7th MED GRP-DYESS	112	6.98	0.97	4.97	8.99
62nd MED GRP-MCCHORD	395	3.06	0.46	2.10	4.02
WALTER REED AMC-WASHINGTON DC	37	26.37	3.07	19.99	32.75
305th MED GRP-MCGUIRE	326	5.84	2.11	1.45	10.24
314th MED GRP-LITTLE ROCK	13	5.71	0.70	4.25	7.17
4th MED GRP-SEYMOUR JOHNSON	90	4.72	0.82	3.02	6.42
92nd MED GRP-FAIRCHILD	128	5.55	0.68	4.14	6.96
FOX AHC-REDSTONE ARSENAL	1	5.73	0.85	3.97	7.49
NH TWENTYNINE PALMS	30	7.24	0.85	5.47	9.02
436th MED GRP-DOVER	36	4.87	0.64	3.54	6.21
NH BEAUFORT	104	6.21	0.85	4.44	7.98
20th MED GRP-SHAW	101	4.91	0.79	3.27	6.55
82nd MED GRP-SHEPPARD	113	5.48	0.64	4.15	6.80
LYSTER AHC-FT. RUCKER	3	6.61	0.65	5.26	7.96
2nd MED GRP-BARKSDALE	62	7.29	0.94	5.34	9.24
325th MED GRP-TYNDALL	43	6.63	0.97	4.61	8.64
KENNER AHC-FT. LEE	122	4.88	0.56	3.73	6.03

Table B.2—Continued

MTF Name	Parent DMIS ID	Mean	Standard Deviation	95% Lower Bound	95% Upper Bound
R W BLISS AHC-FT. HUACHUCA	8	5.48	0.61	4.22	6.74
NH CHARLESTON	103	8.72	0.99	6.66	10.78
NH CORPUS CHRISTI	118	5.75	0.71	4.28	7.23
MUNSON AHC-FT. LEAVENWORTH	58	4.92	0.58	3.72	6.12
NH OAK HARBOR	127	5.96	0.40	5.13	6.79
NH LEMOORE	28	6.28	0.81	4.58	7.97
45th MED GRP-PATRICK	46	5.49	0.47	4.50	6.47
BAYNE-JONES ACH-FT. POLK	64	6.47	0.66	5.10	7.84
NH CHERRY POINT	92	6.31	0.67	4.92	7.70
75th MED GRP-HILL	119	5.49	0.43	4.59	6.39
NHC GREAT LAKES	56	6.07	0.81	4.38	7.75
42ND MEDICAL GROUP-MAXWELL	4	5.56	0.79	3.92	7.20
GUTHRIE AHC-FT. DRUM	330	5.03	1.06	2.84	7.23
78th MED GRP-ROBINS	51	5.35	0.39	4.53	6.17
377th MED GRP-KIRTLAND	83	2.84	0.60	1.60	4.07
L. WOOD ACH-FT. LEONARD WOOD	75	8.55	1.06	6.35	10.76
21st MED GRP-PETERSON	252	4.67	0.50	3.63	5.72
72nd MED GRP-TINKER	96	6.34	0.35	5.61	7.08
IRWIN ACH-FT. RILEY	57	7.07	0.98	5.04	9.10
MONCRIEF ACH-FT. JACKSON	105	5.35	0.52	4.28	6.43
12th MED GRP-RANDOLPH	366	2.07	0.27	1.52	2.63
NHC QUANTICO	385	3.85	0.82	2.16	5.55
355th MED GRP-DAVIS MONTHAN	10	5.77	0.59	4.55	6.99
10th MED GROUP-USAF ACADEMY CO	33	5.05	0.75	3.49	6.60
IRELAND ACH-FT. KNOX	61	6.85	1.15	4.46	9.25

Table B.2—Continued

MTF Name	Parent DMIS ID	Mean	Standard Deviation	95% Lower Bound	95% Upper Bound
56th MED GRP-LUKE	9	7.25	0.60	6.00	8.50
55th MED GRP-OFFUTT	78	6.33	0.82	4.62	8.04
REYNOLDS ACH-FT. SILL	98	7.32	0.47	6.35	8.29
NAVAL HEALTH CARE NEW ENGLAND	6223	6.57	1.04	4.42	8.73
NHC HAWAII	280	4.63	0.37	3.86	5.39
1st MED GRP-LANGLEY	120	4.07	0.22	3.62	4.52
MCDONALD AHC-FT. EUSTIS	121	4.33	0.40	3.50	5.17
81st MED GRP-KEESLER	73	8.29	0.77	6.69	9.89
375th MED GRP-SCOTT	55	6.15	0.38	5.37	6.94
79th MED GRP-ANDREWS	66	4.84	0.34	4.13	5.55
3rd MED GRP-ELMENDORF	6	5.59	0.31	4.95	6.24
NNMC BETHESDA	67	9.83	1.20	7.33	12.33
6th MED GRP-MACDILL	45	5.53	0.69	4.11	6.96
96th MED GRP-EGLIN	42	6.30	0.85	4.53	8.07
88th MED GRP-WRIGHT-PATTERSON	95	6.32	0.49	5.29	7.35
WILLIAM BEAUMONT AMC-FT. BLISS	108	7.57	0.86	5.77	9.36
60th MED GRP-TRAVIS	14	6.86	0.66	5.48	8.24
99th MED GRP-O'CALLAGHAN HOSP	79	6.89	0.73	5.39	8.40
NH CAMP LEJEUNE	91	6.96	0.58	5.75	8.18
EVANS ACH-FT. CARSON	32	7.20	0.78	5.58	8.82
MARTIN ACH-FT. BENNING	48	5.74	0.50	4.70	6.78
NH BREMERTON	126	5.12	0.66	3.76	6.49
KIMBROUGH AMB CAR CEN-FT. MEADE	69	4.34	0.50	3.31	5.37
TRIPLER AMC-FT. SHAFTER	52	8.17	0.84	6.42	9.92
BROOKE AMC-FT. SAM HOUSTON	109	10.01	0.71	8.52	11.49

Table B.2—Continued

MTF Name	Parent DMIS ID	Mean	Standard Deviation	95% Lower Bound	95% Upper Bound
WINN ACH-FT. STEWART	49	5.64	0.61	4.36	6.91
NH PENSACOLA	38	5.90	0.43	5.01	6.78
EISENHOWER AMC-FT. GORDON	47	6.16	0.46	5.20	7.12
59th MED WING-LACKLAND	117	9.59	0.74	8.04	11.13
BLANCHFIELD ACH-FT. CAMPBELL	60	6.92	1.16	4.50	9.34
NH JACKSONVILLE	39	6.04	0.47	5.07	7.02
NH CAMP PENDLETON	24	5.96	0.48	4.96	6.96
MADIGAN AMC-FT. LEWIS	125	7.50	0.66	6.13	8.87
DARNALL AMC-FT. HOOD	110	6.18	0.52	5.08	7.27
WOMACK AMC-FT. BRAGG	89	7.09	0.84	5.34	8.84
DEWITT ACH-FT. BELVOIR	123	3.16	0.34	2.45	3.86
NMC SAN DIEGO	29	8.24	0.74	6.69	9.78
NMC PORTSMOUTH	124	7.77	1.12	5.45	10.10

NOTE: These outcomes are measured at a quarterly frequency and averaged over FYs 2004–2005.

Table B.3
Monthly Outpatient Utilization (Relative Value Units) per 100 Equivalent Lives, FYs 2004–2005

MTF Name	Parent DMIS ID	Mean	Standard Deviation	95% Lower Bound	95% Upper Bound
460th MED GRP-BUCKLEY AFB	7200	60.6	7.0	44.8	76.3
71st MED GRP-VANCE	338	70.4	6.0	57.0	83.8
47th MED GRP-LAUGHLIN	114	53.7	8.0	35.7	71.7
PATTERSON AHC-FT. MONMOUTH	81	79.5	3.8	70.9	88.0
14th MED GRP-COLUMBUS	74	56.3	6.8	41.1	71.5
97th MED GRP-ALTUS	97	64.3	5.1	52.9	75.7
17th MED GRP-GOODFELLOW	364	71.9	9.2	51.2	92.6
319th MED GRP-GRAND FORKS	93	59.1	8.2	40.7	77.6
354th MED GRP-EIELSON	203	58.7	3.2	51.6	65.9
NHC ANNAPOLIS	306	65.8	3.1	58.8	72.8
66th MED GRP-HANSCOM	310	67.2	3.7	58.9	75.5
61st MED SQUAD-LOS ANGELES	248	60.0	3.0	53.2	66.8
9th MED GRP-BEALE	15	58.5	3.2	51.3	65.8
579th MED GROUP-BOLLING	413	58.0	4.3	48.3	67.6
KELLER ACH-WEST POINT	86	66.1	5.1	54.7	77.6
30th MED GRP-VANDENBERG	18	65.2	5.8	52.2	78.1
27th MED GRP-CANNON	85	59.8	4.8	49.1	70.6
341st MED GRP-MALMSTROM	77	66.0	5.3	54.1	77.9
5th MED GRP-MINOT	94	56.9	4.6	46.7	67.2
90th MED GRP-F.E. WARREN	129	68.2	5.0	57.1	79.4
WEED ACH-FT. IRWIN	131	63.1	6.5	48.6	77.6
95th MED GRP-EDWARDS	19	59.1	4.2	49.7	68.6
23rd MED GRP-MOODY	50	68.3	3.7	60.0	76.6
49th MED GRP-HOLLOMAN	84	58.3	4.5	48.3	68.4
437th MED GRP-CHARLESTON	356	59.6	4.5	49.4	69.7

Table B.3—Continued

MTF Name	Parent DMIS ID	Mean	Standard Deviation	95% Lower Bound	95% Upper Bound
366th MED GRP-MOUNTAIN HOME	53	59.7	3.5	51.8	67.7
43rd MEDICAL GROUP-POPE	335	56.2	4.5	46.2	66.2
NHC PATUXENT RIVER	68	69.1	3.2	62.0	76.3
509th MED GRP-WHITEMAN	76	65.5	5.5	53.0	77.9
28th MED GRP-ELLSWORTH	106	61.8	3.8	53.4	70.3
22nd MED GRP-MCCONNELL	59	71.8	4.2	62.3	81.3
BASSETT ACH-FT. WAINWRIGHT	5	70.0	6.0	56.5	83.5
15th MED GRP-HICKAM	287	62.8	4.9	51.7	73.9
1st SPEC OPS MED GRP-HURLBURT	7139	65.6	5.7	52.9	78.3
7th MED GRP-DYESS	112	69.8	8.4	51.1	88.6
62nd MED GRP-MCCHORD	395	61.9	3.5	54.0	69.8
WALTER REED AMC-WASHINGTON DC	37	114.9	18.2	74.1	155.7
305th MED GRP-MCGUIRE	326	59.2	4.9	48.3	70.1
314th MED GRP-LITTLE ROCK	13	67.3	5.0	56.1	78.5
4th MED GRP-SEYMOUR JOHNSON	90	59.8	4.5	49.8	69.8
92nd MED GRP-FAIRCHILD	128	60.4	5.8	47.2	73.5
FOX AHC-REDSTONE ARSENAL	1	62.6	3.5	54.7	70.6
NH TWENTYNINE PALMS	30	72.1	6.1	58.5	85.8
436th MED GRP-DOVER	36	60.2	3.9	51.5	68.9
NH BEAUFORT	104	60.5	4.7	50.0	70.9
20th MED GRP-SHAW	101	60.3	6.2	46.4	74.3
82nd MED GRP-SHEPPARD	113	75.3	4.2	65.8	84.8
LYSTER AHC-FT. RUCKER	3	63.5	7.1	47.5	79.5
2nd MED GRP-BARKSDALE	62	62.4	4.2	53.0	71.8
325th MED GRP-TYNDALL	43	70.0	4.4	60.2	79.8
KENNER AHC-FT. LEE	122	57.2	2.1	52.6	61.9

Table B.3—Continued

MTF Name	Parent DMIS ID	Mean	Standard Deviation	95% Lower Bound	95% Upper Bound
R W BLISS AHC-FT. HUACHUCA	8	58.4	1.7	54.4	62.3
NH CHARLESTON	103	61.2	5.1	49.7	72.7
NH CORPUS CHRISTI	118	65.7	2.8	59.3	72.0
MUNSON AHC-FT. LEAVENWORTH	58	71.0	4.2	61.6	80.5
NH OAK HARBOR	127	68.3	5.3	56.4	80.2
NH LEMOORE	28	61.5	3.5	53.5	69.4
45th MED GRP-PATRICK	46	66.2	5.2	54.4	78.0
BAYNE-JONES ACH-FT. POLK	64	73.4	4.5	63.4	83.4
NH CHERRY POINT	92	68.0	3.6	60.0	76.0
75th MED GRP-HILL	119	67.7	3.1	60.8	74.7
NHC GREAT LAKES	56	68.3	3.7	60.0	76.7
42nd MEDICAL GROUP-MAXWELL	4	67.5	3.1	60.4	74.5
GUTHRIE AHC-FT. DRUM	330	76.2	2.8	70.0	82.4
78th MED GRP-ROBINS	51	64.7	3.0	58.1	71.3
377th MED GRP-KIRTLAND	83	46.5	4.5	36.3	56.7
L. WOOD ACH-FT. LEONARD WOOD	75	67.0	3.5	59.2	74.9
21st MED GRP-PETERSON	252	64.8	2.9	58.3	71.3
72nd MED GRP-TINKER	96	66.1	2.4	60.7	71.6
IRWIN ACH-FT. RILEY	57	69.2	4.1	60.0	78.3
MONCRIEF ACH-FT. JACKSON	105	60.4	2.6	54.7	66.2
12th MED GRP-RANDOLPH	366	54.3	2.0	49.9	58.7
NHC QUANTICO	385	64.6	3.3	57.2	72.0
355th MED GRP-DAVIS MONTHAN	10	59.2	3.4	51.7	66.8
10th MED GROUP-USAF ACADEMY CO	33	57.5	3.1	50.6	64.4
IRELAND ACH-FT. KNOX	61	70.4	3.7	62.2	78.6

Table B.3—Continued

MTF Name	Parent DMIS ID	Mean	Standard Deviation	95% Lower Bound	95% Upper Bound
56th MED GRP-LUKE	9	58.0	6.6	43.2	72.8
55th MED GRP-OFFUTT	78	64.4	6.6	49.7	79.2
REYNOLDS ACH-FT. SILL	98	67.8	4.7	57.3	78.3
NAVAL HEALTH CARE NEW ENGLAND	6223	75.7	4.9	64.6	86.8
NHC HAWAII	280	76.2	3.3	68.8	83.5
1st MED GRP-LANGLEY	120	64.0	6.0	50.4	77.5
MCDONALD AHC-FT. EUSTIS	121	64.0	4.3	54.4	73.6
81st MED GRP-KEESLER	73	59.5	5.5	47.1	71.9
375th MED GRP-SCOTT	55	58.3	2.1	53.6	63.0
79th MED GRP-ANDREWS	66	60.6	2.6	54.8	66.4
3rd MED GRP-ELMENDORF	6	50.9	3.2	43.8	58.0
NNMC BETHESDA	67	74.6	4.6	64.3	84.9
6th MED GRP-MACDILL	45	54.4	4.7	43.9	65.0
96th MED GRP-EGLIN	42	61.6	3.7	53.3	69.8
88th MED GRP-WRIGHT-PATTERSON	95	59.2	5.7	46.4	72.1
WILLIAM BEAUMONT AMC-FT. BLISS	108	51.6	3.8	43.0	60.2
60th MED GRP-TRAVIS	14	48.0	4.6	37.7	58.3
99th MED GRP-O'CALLAGHAN HOSP	79	54.1	5.1	42.7	65.6
NH CAMP LEJEUNE	91	68.1	5.7	55.2	80.9
EVANS ACH-FT. CARSON	32	69.3	1.7	65.5	73.1
MARTIN ACH-FT. BENNING	48	63.6	4.9	52.6	74.6
NH BREMERTON	126	61.8	2.8	55.5	68.1
KIMBROUGH AMB CAR CEN-FT. MEADE	69	67.3	2.6	61.5	73.1
TRIPLER AMC-FT. SHAFTER	52	78.6	3.3	71.1	86.1
BROOKE AMC-FT. SAM HOUSTON	109	66.9	2.0	62.3	71.5

Table B.3—Continued

MTF Name	Parent DMIS ID	Mean	Standard Deviation	95% Lower Bound	95% Upper Bound
WINN ACH-FT. STEWART	49	65.1	4.4	55.2	75.0
NH PENSACOLA	38	62.7	3.8	54.2	71.2
EISENHOWER AMC-FT. GORDON	47	57.5	2.2	52.7	62.3
59th MED WING-LACKLAND	117	57.9	2.8	51.7	64.1
BLANCHFIELD ACH-FT. CAMPBELL	60	63.5	3.6	55.4	71.6
NH JACKSONVILLE	39	71.8	2.7	65.8	77.8
NH CAMP PENDLETON	24	55.1	3.0	48.3	61.9
MADIGAN AMC-FT. LEWIS	125	65.5	1.6	61.8	69.2
DARNALL AMC-FT. HOOD	110	72.1	4.6	61.7	82.5
WOMACK AMC-FT. BRAGG	89	62.9	3.8	54.2	71.5
DEWITT ACH-FT. BELVOIR	123	61.6	2.0	57.1	66.1
NMC SAN DIEGO	29	55.8	2.2	50.9	60.7
NMC PORTSMOUTH	124	69.1	2.8	62.9	75.3

NOTE: These outcomes are measured at a quarterly frequency and averaged over FYs 2004–2005.

Table B.4
Monthly Drug Utilization (30-Day Prescriptions Filled) per 100 Equivalent
Lives, FYs 2004–2005

MTF Name	Parent DMIS ID	Mean	Standard Deviation	95% Lower Bound	95% Upper Bound
460th MED GRP-BUCKLEY AFB	7200	66.3	5.1	55.1	77.6
71st MED GRP-VANCE	338	123.7	6.0	110.4	137.0
47th MED GRP-LAUGHLIN	114	101.6	5.0	90.7	112.6
PATTERSON AHC-FT. MONMOUTH	81	88.9	4.3	79.4	98.4
14th MED GRP-COLUMBUS	74	106.8	7.2	90.9	122.8
97th MED GRP-ALTUS	97	107.0	6.6	92.4	121.6
17th MED GRP-GOODFELLOW	364	100.1	4.3	90.5	109.6
319th MED GRP-GRAND FORKS	93	77.9	5.4	65.9	89.9
354th MED GRP-EIELSON	203	65.8	2.8	59.7	71.9
NHC ANNAPOLIS	306	121.6	7.3	105.4	137.8
66th MED GRP-HANSCOM	310	83.6	2.5	78.0	89.1
61st MED SQUAD-LOS ANGELES	248	66.3	3.8	58.0	74.6
9th MED GRP-BEALE	15	83.0	4.5	73.2	92.9
579th MED GROUP-BOLLING	413	77.1	4.8	66.5	87.8
KELLER ACH-WEST POINT	86	93.9	5.9	80.9	106.9
30th MED GRP-VANDENBERG	18	77.6	2.5	72.1	83.1
27th MED GRP-CANNON	85	91.3	3.2	84.3	98.4
341st MED GRP-MALMSTROM	77	83.2	5.3	71.4	95.0
5th MED GRP-MINOT	94	90.4	3.0	83.8	96.9
90th MED GRP-F.E. WARREN	129	85.9	3.6	77.9	94.0
WEED ACH-FT. IRWIN	131	79.1	6.4	64.9	93.2
95th MED GRP-EDWARDS	19	71.3	3.8	62.8	79.8
23rd MED GRP-MOODY	50	90.1	7.4	73.7	106.4
49th MED GRP-HOLLOMAN	84	104.3	4.2	95.1	113.5
437th MED GRP-CHARLESTON	356	96.8	2.9	90.4	103.2

Table B.4—Continued

MTF Name	Parent DMIS ID	Mean	Standard Deviation	95% Lower Bound	95% Upper Bound
366th MED GRP-MOUNTAIN HOME	53	87.8	4.6	77.5	98.0
43rd MEDICAL GROUP-POPE	335	80.9	2.5	75.4	86.3
NHC PATUXENT RIVER	68	105.0	3.4	97.6	112.5
509th MED GRP-WHITEMAN	76	98.7	6.6	84.1	113.4
28th MED GRP-ELLSWORTH	106	93.8	2.5	88.4	99.2
22nd MED GRP-MCCONNELL	59	126.9	5.5	114.8	139.0
BASSETT ACH-FT. WAINWRIGHT	5	90.4	2.8	84.2	96.7
15th MED GRP-HICKAM	287	78.3	7.0	62.9	93.7
1st SPEC OPS MED GRP-HURLBURT	7139	85.2	5.2	73.8	96.6
7th MED GRP-DYESS	112	101.8	7.0	86.4	117.2
62nd MED GRP-MCCHORD	395	93.0	3.4	85.4	100.6
WALTER REED AMC-WASHINGTON DC	37	143.4	5.9	130.5	156.4
305th MED GRP-MCGUIRE	326	74.0	3.0	67.4	80.6
314th MED GRP-LITTLE ROCK	13	97.9	4.6	87.7	108.1
4th MED GRP-SEYMOUR JOHNSON	90	93.3	4.8	82.7	104.0
92nd MED GRP-FAIRCHILD	128	88.3	3.6	80.3	96.2
FOX AHC-REDSTONE ARSENAL	1	108.5	2.6	102.7	114.3
NH TWENTYNINE PALMS	30	88.4	3.7	80.1	96.6
436th MED GRP-DOVER	36	95.8	4.3	86.3	105.3
NH BEAUFORT	104	99.9	5.2	88.5	111.4
20th MED GRP-SHAW	101	93.0	4.1	84.0	102.0
82nd MED GRP-SHEPPARD	113	113.6	12.1	86.8	140.4
LYSTER AHC-FT. RUCKER	3	120.1	7.5	103.4	136.7
2nd MED GRP-BARKSDALE	62	97.2	3.0	90.7	103.8
325th MED GRP-TYNDALL	43	104.4	4.2	95.2	113.6
KENNER AHC-FT. LEE	122	87.4	3.5	79.8	95.1

Table B.4—Continued

MTF Name	Parent DMIS ID	Mean	Standard Deviation	95% Lower Bound	95% Upper Bound
R W BLISS AHC-FT. HUACHUCA	8	94.8	3.0	88.1	101.5
NH CHARLESTON	103	113.7	3.1	106.8	120.6
NH CORPUS CHRISTI	118	108.3	4.3	98.8	117.8
MUNSON AHC-FT. LEAVENWORTH	58	103.4	5.9	90.3	116.5
NH OAK HARBOR	127	95.4	4.0	86.7	104.1
NH LEMOORE	28	102.3	5.7	89.8	114.8
45th MED GRP-PATRICK	46	110.9	5.7	98.3	123.6
BAYNE-JONES ACH-FT. POLK	64	104.3	4.6	94.2	114.4
NH CHERRY POINT	92	98.5	2.0	94.0	103.0
75th MED GRP-HILL	119	91.5	3.2	84.5	98.5
NHC GREAT LAKES	56	105.4	3.2	98.4	112.4
42nd MEDICAL GROUP-MAXWELL	4	104.1	3.6	96.2	112.1
GUTHRIE AHC-FT. DRUM	330	93.6	4.5	83.6	103.6
78th MED GRP-ROBINS	51	112.6	2.9	106.2	119.1
377th MED GRP-KIRTLAND	83	94.8	4.7	84.4	105.2
L. WOOD ACH-FT. LEONARD WOOD	75	125.0	6.6	110.6	139.5
21st MED GRP-PETERSON	252	90.3	4.6	80.0	100.6
72nd MED GRP-TINKER	96	103.8	6.1	90.4	117.3
IRWIN ACH-FT. RILEY	57	97.9	7.3	81.8	114.1
MONCRIEF ACH-FT. JACKSON	105	112.1	2.2	107.2	117.0
12th MED GRP-RANDOLPH	366	111.2	4.0	102.4	120.1
NHC QUANTICO	385	90.8	4.2	81.6	100.0
355th MED GRP-DAVIS MONTHAN	10	94.3	3.6	86.3	102.2
10th MED GROUP-USAF ACADEMY CO	33	99.1	6.5	84.7	113.5
IRELAND ACH-FT. KNOX	61	127.6	4.7	117.1	138.0

Table B.4—Continued

MTF Name	Parent DMIS ID	Mean	Standard Deviation	95% Lower Bound	95% Upper Bound
56th MED GRP-LUKE	9	98.1	7.5	81.5	114.7
55th MED GRP-OFFUTT	78	94.9	7.3	78.8	110.9
REYNOLDS ACH-FT. SILL	98	93.9	3.6	86.0	101.9
NAVAL HEALTH CARE NEW ENGLAND	6223	96.9	3.7	88.8	105.0
NHC HAWAII	280	84.3	5.1	73.0	95.7
1st MED GRP-LANGLEY	120	90.6	5.8	77.7	103.5
MCDONALD AHC-FT. EUSTIS	121	104.9	4.4	95.2	114.6
81st MED GRP-KEESLER	73	140.1	11.5	114.6	165.6
375th MED GRP-SCOTT	55	105.2	3.3	97.9	112.4
79th MED GRP-ANDREWS	66	91.3	4.8	80.6	102.0
3rd MED GRP-ELMENDORF	6	82.9	2.6	77.3	88.5
NNMC BETHESDA	67	116.4	4.2	107.1	125.8
6th MED GRP-MACDILL	45	101.8	6.0	88.4	115.1
96th MED GRP-EGLIN	42	108.0	4.9	97.1	118.8
88th MED GRP-WRIGHT-PATTERSON	95	132.2	4.4	122.5	141.8
WILLIAM BEAUMONT AMC-FT. BLISS	108	110.8	5.8	98.1	123.6
60th MED GRP-TRAVIS	14	99.6	5.2	88.0	111.1
99th MED GRP-O'CALLAGHAN HOSP	79	102.8	5.6	90.5	115.0
NH CAMP LEJEUNE	91	96.5	2.4	91.1	101.9
EVANS ACH-FT. CARSON	32	93.6	2.3	88.4	98.8
MARTIN ACH-FT. BENNING	48	106.9	7.8	89.8	124.1
NH BREMERTON	126	111.3	2.7	105.3	117.4
KIMBROUGH AMB CAR CEN-FT. MEADE	69	105.7	5.1	94.6	116.9
TRIPLER AMC-FT. SHAFTER	52	87.7	4.8	77.1	98.4
BROOKE AMC-FT. SAM HOUSTON	109	139.7	3.6	131.8	147.6

Table B.4—Continued

MTF Name	Parent DMIS ID	Mean	Standard Deviation	95% Lower Bound	95% Upper Bound
WINN ACH-FT. STEWART	49	101.5	4.0	92.7	110.3
NH PENSACOLA	38	112.6	4.3	103.0	122.2
EISENHOWER AMC-FT. GORDON	47	128.9	4.5	119.1	138.7
59th MED WING-LACKLAND	117	119.4	2.5	113.8	125.0
BLANCHFIELD ACH-FT. CAMPBELL	60	99.7	3.6	91.8	107.7
NH JACKSONVILLE	39	107.9	3.5	100.1	115.6
NH CAMP PENDLETON	24	79.8	3.3	72.5	87.1
MADIGAN AMC-FT. LEWIS	125	106.9	4.9	96.1	117.6
DARNALL AMC-FT. HOOD	110	97.6	7.7	80.7	114.6
WOMACK AMC-FT. BRAGG	89	87.5	4.8	76.8	98.1
DEWITT ACH-FT. BELVOIR	123	100.6	3.8	92.3	109.0
NMC SAN DIEGO	29	97.0	3.2	90.0	104.1
NMC PORTSMOUTH	124	114.5	4.4	104.6	124.3

NOTE: These outcomes are measured at a quarterly frequency and averaged over FYs 2004–2005.

Table B.5
Annual Total Cost Per Equivalent Life, FYs 2004–2005

MTF Name	Parent DMIS ID	Mean ($)	Standard Deviation ($)	95% Lower Bound ($)	95% Upper Bound ($)
460th MED GRP-BUCKLEY AFB	7200	2,179	395	1,299	3,059
71st MED GRP-VANCE	338	2,578	215	2,099	3,057
47th MED GRP-LAUGHLIN	114	2,185	184	1,775	2,595
PATTERSON AHC-FT. MONMOUTH	81	2,581	229	2,070	3,092
14th MED GRP-COLUMBUS	74	1,943	169	1,567	2,320
97th MED GRP-ALTUS	97	2,633	120	2,366	2,900
17th MED GRP-GOODFELLOW	364	2,177	239	1,644	2,710
319th MED GRP-GRAND FORKS	93	2,338	356	1,546	3,130
354th MED GRP-EIELSON	203	2,443	236	1,917	2,968
NHC ANNAPOLIS	306	2,124	123	1,849	2,398
66th MED GRP-HANSCOM	310	2,444	147	2,117	2,770
61st MED SQUAD-LOS ANGELES	248	2,154	263	1,568	2,740
9th MED GRP-BEALE	15	2,341	244	1,799	2,884
579th MED GROUP-BOLLING	413	2,179	178	1,783	2,575
KELLER ACH-WEST POINT	86	2,684	180	2,285	3,084
30th MED GRP-VANDENBERG	18	2,127	240	1,593	2,662
27th MED GRP-CANNON	85	2,604	234	2,082	3,125
341st MED GRP-MALMSTROM	77	2,259	139	1,949	2,568
5th MED GRP-MINOT	94	2,228	96	2,014	2,441
90th MED GRP-F.E. WARREN	129	2,563	221	2,071	3,054
WEED ACH-FT. IRWIN	131	2,868	340	2,111	3,626
95th MED GRP-EDWARDS	19	2,201	208	1,738	2,663
23rd MED GRP-MOODY	50	2,211	100	1,990	2,433
49th MED GRP-HOLLOMAN	84	2,672	671	1,178	4,166
437th MED GRP-CHARLESTON	356	2,040	138	1,732	2,347

Table B.5—Continued

MTF Name	Parent DMIS ID	Mean ($)	Standard Deviation ($)	95% Lower Bound ($)	95% Upper Bound ($)
366th MED GRP-MOUNTAIN HOME	53	2,462	149	2,131	2,793
43rd MEDICAL GROUP-POPE	335	1,822	98	1,603	2,041
NHC PATUXENT RIVER	68	2,406	90	2,206	2,607
509th MED GRP-WHITEMAN	76	2,274	150	1,940	2,608
28th MED GRP-ELLSWORTH	106	2,406	484	1,329	3,483
22nd MED GRP-MCCONNELL	59	2,246	129	1,959	2,534
BASSETT ACH-FT. WAINWRIGHT	5	2,958	244	2,414	3,501
15th MED GRP-HICKAM	287	1,952	177	1,557	2,347
1st SPEC OPS MED GRP-HURLBURT	7139	2,000	143	1,682	2,318
7th MED GRP-DYESS	112	2,436	230	1,923	2,948
62nd MED GRP-MCCHORD	395	1,764	136	1,460	2,067
WALTER REED AMC-WASHINGTON DC	37	6,904	382	6,053	7,755
305th MED GRP-MCGUIRE	326	2,413	367	1,595	3,230
314th MED GRP-LITTLE ROCK	13	2,008	104	1,776	2,240
4th MED GRP-SEYMOUR JOHNSON	90	1,950	141	1,636	2,263
92nd MED GRP-FAIRCHILD	128	2,078	147	1,751	2,406
FOX AHC-REDSTONE ARSENAL	1	2,225	137	1,920	2,530
NH TWENTYNINE PALMS	30	3,076	187	2,659	3,493
436th MED GRP-DOVER	36	1,807	206	1,347	2,266
NH BEAUFORT	104	2,443	282	1,817	3,070
20th MED GRP-SHAW	101	1,903	116	1,646	2,161
82nd MED GRP-SHEPPARD	113	2,477	192	2,049	2,905
LYSTER AHC-FT. RUCKER	3	2,163	120	1,896	2,430
2nd MED GRP-BARKSDALE	62	2,198	205	1,742	2,655
325th MED GRP-TYNDALL	43	2,288	143	1,970	2,607

Table B.5—Continued

MTF Name	Parent DMIS ID	Mean ($)	Standard Deviation ($)	95% Lower Bound ($)	95% Upper Bound ($)
KENNER AHC-FT. LEE	122	1,907	190	1,483	2,331
R W BLISS AHC-FT. HUACHUCA	8	2,266	204	1,812	2,720
NH CHARLESTON	103	2,761	269	2,161	3,360
NH CORPUS CHRISTI	118	2,418	128	2,133	2,704
MUNSON AHC-FT. LEAVENWORTH	58	2,326	162	1,965	2,686
NH OAK HARBOR	127	2,191	88	1,994	2,387
NH LEMOORE	28	2,590	155	2,245	2,936
45th MED GRP-PATRICK	46	2,069	108	1,829	2,309
BAYNE-JONES ACH-FT. POLK	64	2,407	74	2,242	2,571
NH CHERRY POINT	92	2,379	127	2,095	2,662
75th MED GRP-HILL	119	2,078	98	1,861	2,295
NHC GREAT LAKES	56	2,166	151	1,831	2,502
42ND MEDICAL GROUP-MAXWELL	4	2,095	125	1,816	2,375
GUTHRIE AHC-FT. DRUM	330	1,889	141	1,576	2,203
78th MED GRP-ROBINS	51	2,054	65	1,909	2,199
377th MED GRP-KIRTLAND	83	1,763	148	1,433	2,093
L. WOOD ACH-FT. LEONARD WOOD	75	2,764	177	2,371	3,157
21st MED GRP-PETERSON	252	1,709	131	1,416	2,001
72nd MED GRP-TINKER	96	2,329	58	2,201	2,457
IRWIN ACH-FT. RILEY	57	2,320	258	1,746	2,893
MONCRIEF ACH-FT. JACKSON	105	1,991	112	1,742	2,240
12th MED GRP-RANDOLPH	366	1,814	83	1,630	1,997
NHC QUANTICO	385	2,018	129	1,731	2,304
355th MED GRP-DAVIS MONTHAN	10	2,010	138	1,704	2,317

Table B.5—Continued

MTF Name	Parent DMIS ID	Mean ($)	Standard Deviation ($)	95% Lower Bound ($)	95% Upper Bound ($)
10th MED GROUP-USAF ACADEMY CO	33	2,100	103	1,870	2,330
IRELAND ACH-FT. KNOX	61	2,533	215	2,056	3,011
56th MED GRP-LUKE	9	2,366	249	1,813	2,919
55th MED GRP-OFFUTT	78	2,320	247	1,770	2,871
REYNOLDS ACH-FT. SILL	98	2,147	51	2,035	2,260
NAVAL HEALTH CARE NEW ENGLAND	6223	2,688	296	2,029	3,346
NHC HAWAII	280	2,112	120	1,845	2,379
1st MED GRP-LANGLEY	120	2,054	141	1,740	2,368
MCDONALD AHC-FT. EUSTIS	121	1,968	129	1,681	2,255
81st MED GRP-KEESLER	73	2,653	142	2,336	2,969
375th MED GRP-SCOTT	55	2,434	97	2,218	2,649
79th MED GRP-ANDREWS	66	2,637	171	2,258	3,017
3rd MED GRP-ELMENDORF	6	2,179	95	1,968	2,390
NNMC BETHESDA	67	3,435	284	2,803	4,068
6th MED GRP-MACDILL	45	2,113	158	1,761	2,465
96th MED GRP-EGLIN	42	2,349	97	2,133	2,565
88th MED GRP-WRIGHT-PATTERSON	95	2,834	255	2,267	3,402
WILLIAM BEAUMONT AMC-FT. BLISS	108	2,549	126	2,268	2,831
60th MED GRP-TRAVIS	14	2,558	110	2,314	2,802
99th MED GRP-O'CALLAGHAN HOSP	79	2,647	264	2,061	3,234
NH CAMP LEJEUNE	91	2,299	215	1,821	2,777
EVANS ACH-FT. CARSON	32	2,426	126	2,145	2,706
MARTIN ACH-FT. BENNING	48	1,964	102	1,736	2,191
NH BREMERTON	126	2,497	171	2,117	2,877

Table B.5—Continued

MTF Name	Parent DMIS ID	Mean ($)	Standard Deviation ($)	95% Lower Bound ($)	95% Upper Bound ($)
KIMBROUGH AMB CAR CEN-FT. MEADE	69	2,293	142	1,976	2,609
TRIPLER AMC-FT. SHAFTER	52	2,902	200	2,456	3,348
BROOKE AMC-FT. SAM HOUSTON	109	3,215	134	2,918	3,512
WINN ACH-FT. STEWART	49	2,146	288	1,504	2,787
NH PENSACOLA	38	2,183	80	2,005	2,360
EISENHOWER AMC-FT. GORDON	47	2,394	228	1,887	2,901
59th MED WING-LACKLAND	117	3,093	137	2,787	3,399
BLANCHFIELD ACH-FT. CAMPBELL	60	2,142	124	1,865	2,418
NH JACKSONVILLE	39	2,365	93	2,158	2,573
NH CAMP PENDLETON	24	2,235	151	1,898	2,572
MADIGAN AMC-FT. LEWIS	125	2,454	224	1,955	2,953
DARNALL AMC-FT. HOOD	110	2,074	131	1,781	2,366
WOMACK AMC-FT. BRAGG	89	1,992	184	1,583	2,401
DEWITT ACH-FT. BELVOIR	123	2,059	132	1,765	2,352
NMC SAN DIEGO	29	2,423	159	2,070	2,777

NOTE: These outcomes are measured at a quarterly frequency and averaged over FYs 2004–2005.

Table B.6
Standard Deviations of MTF Outcomes in FYs 2004–2005 and FY 2006,
Mean by MTF Size Quintile

Size Quintile	Standard Deviation, FYs 2004–2005	Standard Deviation, FY 2006	Ratio of FY 2006 to FYs 2004–2005
Inpatient utilization			
All	0.57	0.66	1.61
1 (smallest)	0.82	0.92	2.33
2	0.68	0.76	1.49
3	0.49	0.70	1.65
4	0.43	0.46	1.30
5 (largest)	0.40	0.44	1.27
Outpatient utilization			
All	3.2	4.3	1.47
1 (smallest)	3.4	4.9	1.53
2	3.7	5.1	1.40
3	3.0	3.9	1.41
4	3.4	4.0	1.30
5 (largest)	2.4	3.7	1.74
Drug utilization			
All	3.4	4.4	1.48
1 (smallest)	3.4	4.3	1.43
2	3.3	4.3	1.56
3	3.3	4.8	1.66
4	3.9	4.7	1.32
5 (largest)	3.0	3.9	1.43
Total cost			
All	$112	$147	1.75
1 (smallest)	$141	$230	2.04
2	$146	$147	1.56
3	$94	$119	1.65
4	$91	$109	1.55
5 (largest)	$87	$129	1.94

NOTE: Standard deviations are in levels. Size quintiles are defined by mean non-
active duty Prime enrollment in FYs 2004–2005; see Table 4.3.

Outpatient Utilization and MTF Size

Mean outpatient utilization among MTFs' enrollees decreased (albeit modestly) with MTF size in Figure 4.3, while the mean levels of the other outcomes increased with size. OASD(HA) has noted that outpatient care received from MTFs does not include ancillary workload, while outpatient care received from civilian health care providers does. If large MTFs are less reliant on civilian providers for this care, mean outpatient utilization may appear to be lower at these MTFs, even if outpatient utilization did not vary (or actually increased) with size. We investigate this possibility by regressing mean outpatient utilization at MTFs during FYs 2004–2005 on the number of non-active duty enrollees, as well as the share of outpatient utilization delivered by civilian providers.[1]

The results appear in Table C.1. In the first specification, a larger number of non-active duty enrollees is associated with modestly lower utilization; we cannot, however, reject the hypothesis that there is in fact no relationship. In the second specification, the civilian share of outpatient care is included. We cannot reject the hypothesis that this share is unrelated to utilization, suggesting that under-reporting of outpatient workload at MTFs does not lead to a spurious relationship between mean outpatient utilization and MTF size. Indeed, we are again unable to reject the hypothesis that utilization is unrelated to the number of non-active duty enrollees, suggesting that this relationship is negligible.

---

[1] The results are similar when size is measured by the number of non-active duty equivalent lives enrolled at MTFs.

Table C.1
Regression of Outpatient Utilization

	Specification	
	1	2
Parameter Estimate (Standard Error)		
Constant	64.668***	61.146***
	(1.157)	(2.767)
Number of non-active duty enrollees (thousands)	−0.029	−0.064
	(0.048)	(0.054)
Civilian share of outpatient utilization (%)	—	0.077
		(0.055)
Other Statistics		
R-squared	0.003	0.020
N		114

NOTES: Dependent and independent variables are FY 2004–2005 means.

* denotes statistical significance at the 10 percent level.

** denotes statistical significance at the 5 percent level.

*** denotes statistical significance at the 1 percent level.

Catastrophic Hospital Admissions

We identified hospital admissions as catastrophic if the DRG associated with the diagnosis was sufficiently resource-intensive. The resource use of DRGs was measured by RWPs. Alternative thresholds were the top 1, 5, and 10 percent of DRGs with respect to RWPs. Table D.1 identifies these catastrophic DRGs, describes them, and notes the most stringent threshold met. For example, DRG 106 (Coronary Bypass with PTCA) is in the top 5 percent of DRGs by RWPs; as a result, this DRG is in the top 10 percent but not the top 1 percent.

Table D.1
Top 10 Percent of DRGs, by RWPs

DRG	Description	Threshold (%)
1	CRANIOTOMY AGE >17 W CC	10
103	HEART & HEART/LUNG TRANSPLANT	1
104	CARDIAC VALVE & OTHER MAJOR CARDIOTHORACIC PROC W CARDIAC CATH	5
105	CARDIAC VALVE & OTHER MAJOR CARDIOTHORACIC PROC W/O CARDIAC CATH	10
106	CORONARY BYPASS W PTCA	5
107	CORONARY BYPASS W CARDIAC CATH	10
108	OTHER CARDIOTHORACIC PROCEDURES	10
109	CORONARY BYPASS W/O CARDIAC CATH	10
110	MAJOR CARDIOVASCULAR PROCEDURES W CC	10
113	AMPUTATION FOR CIRC SYSTEM DISORDERS EXCEPT UPPER LIMB & TOE	10
115	PRM CARD PACEM IMPL W AMI/HF/SHOCK OR AICD LEAD OR GNRTR PROC	10
154	STOMACH: ESOPHAGEAL & DUODENAL PROCEDURES AGE >17 W CC	10
191	PANCREAS: LIVER & SHUNT PROCEDURES W CC	10
201	OTHER HEPATOBILIARY OR PANCREAS O.R. PROCEDURES	10
292	OTHER ENDOCRINE: NUTRIT & METAB O.R. PROC W CC	10
302	KIDNEY TRANSPLANT	10
473	ACUTE LEUKEMIA W/O MAJOR O.R. PROCEDURE AGE >17	5
475	RESPIRATORY SYSTEM DIAGNOSIS WITH VENTILATOR SUPPORT	10
480	LIVER TRANSPLANT	5
481	BONE MARROW TRANSPLANT	5
482	TRACHEOSTOMY FOR FACE, MOUTH & NECK DIAGNOSES	10
483	TRACH W MECH VENT 96+ HRS OR PDX EXCEPT FACE, MOUTH & NECK DIAG	1
484	CRANIOTOMY FOR MULTIPLE SIGNIFICANT TRAUMA	5
485	LIMB REATTACHMENT: HIP AND FEMUR PROC FOR MULTIPLE SIGNIFICANT T	10

Table D.1—Continued

DRG	Description	Threshold (%)
486	OTHER O.R. PROCEDURES FOR MULTIPLE SIGNIFICANT TRAUMA	10
488	HIV W EXTENSIVE O.R. PROCEDURE	10
495	LUNG TRANSPLANT	5
496	COMBINED ANTERIOR/POSTERIOR SPINAL FUSION	10
497	SPINAL FUSION EXCEPT CERVICAL W CC	10
504	EXTENSIVE 3RD DEGREE BURNS W SKIN GRAFT	5
506	FULL THICKNESS BURN W SKIN GRAFT OR INHAL INJ W CC OR SIG TRAUMA	10
512	SIMULTANEOUS PANCREAS/KIDNEY TRANSPLANT	10
515	CARDIAC DEFIBRILLATOR IMPLANT W/O CARDIAC CATH	5
525	HEART ASSIST SYSTEM IMPLANT	1
528	INTRACRANIAL VASCULAR PROCEDURES W PDX HEMORRHAGE	5
531	SPINAL PROCEDURES W CC	10
535	CARDIAC DEFIB IMPLANT W CARDIAC CATH W AMI/HF/SHOCK	5
536	CARDIAC DEFIB IMPLANT W CARDIAC CATH W/O AMI/HF/ SHOCK	5
539	LYMPHOMA & LEUKEMIA W MAJOR O.R. PROCEDURE W CC	10
602	NEONATE: BIRTHWT <750G: DISCHARGED ALIVE	1
603	NEONATE: BIRTHWT <750G: DIED	5
604	NEONATE: BIRTHWT 7500–0999G: DISCHARGED ALIVE	5
606	NEONATE: BIRTHWT 10000–01499G: W SIGNIF OR PROC: DISCHARGED ALIVE	1
607	NEONATE: BIRTHWT 10000–01499G: W/O SIGNIF OR PROC: DISCHARGED ALI	5
608	NEONATE: BIRTHWT 10000–01499G: DIED	5
609	NEONATE: BIRTHWT 15000–01999G: W SIGNIF OR PROC: W MULT MAJOR PROB	5
611	NEONATE: BIRTHWT 15000–01999G: W/O SIGNIF OR PROC: W MULT MAJOR PROB	10
615	NEONATE: BIRTHWT 20000–02499G: W SIGNIF OR PROC: W MULT MAJOR PROB	5

Table D.1—Continued

DRG	Description	Threshold (%)
616	NEONATE: BIRTHWT 20000–02499G: W SIGNIF OR PROC: W/O MULT MAJOR PROB	10
617	NEONATE: BIRTHWT 20000–02499G: W/O SIGNIF OR PROC: W MULT MAJOR PROB	10
622	NEONATE: BIRTHWT >2499G: W SIGNIF OR PROC: W MULT MAJOR PROB	5

Detailed Results of Analysis of Catastrophic Hospital Admissions

Tables E.1–E.3 report the results of the regressions of total inpatient utilization (RWPs) per 1,000 equivalent lives per month on catastrophic utilization—as well as indicator variables for each MTF for the summer, fall, and winter quarters—for FY 2004. The regressions exclude the 460th Medical Group at Buckley Air Force Base (Parent DMIS ID 7200), whose outcomes were extreme outliers. As in our benchmark analysis, the utilization data are quarterly averages. Each table corresponds to a different threshold according to which high-RWP DRGs define catastrophic hospital admissions, as described in Chapter Five.

Table E.1
Results of Regression of Total Inpatient Utilization,
with Catastrophic Admissions Defined by the Highest
1 Percent of DRGs by Resource Use

Parameter Estimate (Standard Error)	
Catastrophic inpatient utilization	1.053*** (0.104)
Summer quarter	0.661*** (0.105)
Fall quarter	0.487*** (0.106)
Winter quarter	0.358*** (0.105)
Other Statistics	
R-squared	0.909
N	452

NOTES: Parameter estimates corresponding to MTF indicator
variables are not reported here.
* denotes statistical significance at the 10 percent level.

** denotes statistical significance at the 5 percent level.

*** denotes statistical significance at the 1 percent level.

Table E.2
Results of Regression of Total Inpatient Utilization,
with Catastrophic Admissions Defined by the Highest
5 Percent of DRGs by Resource Use

Parameter Estimate (Standard Error)	
Catastrophic inpatient utilization	1.013*** (0.064)
Summer quarter	0.528*** (0.092)
Fall quarter	0.364*** (0.092)
Winter quarter	0.249*** (0.091)
Other Statistics	
R-squared	0.932
N	452

NOTES: Parameter estimates corresponding to MTF indicator
variables are not reported here.
* denotes statistical significance at the 10 percent level.

** denotes statistical significance at the 5 percent level.

*** denotes statistical significance at the 1 percent level.

Table E.3
Results of Regression of Total Inpatient Utilization, with Catastrophic Admissions Defined by the Highest 10 Percent of DRGs by Resource Use

	Parameter Estimate (Standard Error)
Catastrophic inpatient utilization	1.028*** (0.050)
Summer quarter	0.456*** (0.080)
Fall quarter	0.272*** (0.081)
Winter quarter	0.174*** (0.080)
Other Statistics	
R-squared	0.948
N	452

NOTES: Parameter estimates corresponding to MTF indicator variables are not reported here.

* denotes statistical significance at the 10 percent level.

** denotes statistical significance at the 5 percent level.

*** denotes statistical significance at the 1 percent level.

Bibliography

Amemiya, Takeshi, *Introduction to Statistics and Econometrics*, Cambridge, Mass.: Harvard University Press, 1994.

Atkinson, Gregory, PMPM Calculation, emailed to the authors, January 3, 2007a.

———, PMPM Summary March 2007 with Beneficiary Categories file, emailed to the authors, May 14, 2007b.

Bronskill, Susan, Sharon-Lise Normand, Mary Beth Landrum, and Robert Rosenheck, "Longitudinal Profiles of Health Care Providers," *Statistics in Medicine*, Vol. 21, 2002, pp. 1067–1088.

Consumer-Purchaser Disclosure Project, *More Efficient Physicians: A Path to Significant Savings in Health Care*, Washington, D.C., 2003.

Donabedian, Avedis, "The Quality of Care: How Can It Be Assessed?" *Journal of the American Medical Association*, Vol. 260, No. 12, 1988, pp. 1743–1748.

Dranove, David, Daniel Kessler, Mark McClellan, and Mark Satterthwaite, "Is More Information Better? The Effects of "Report Cards" on Health Care Providers," *Journal of Political Economy*, Vol. 111, No. 3, 2003, pp. 555–588.

Goldstein, Harvey, and David J. Spiegelhalter, "League Tables and Their Limitations: Statistical Issues in Comparisons of Institutional Performance," *Journal of the Royal Statistical Society, Series A (Statistics in Society)*, Vol. 159, No. 3, 1996, pp. 385–443.

Greene, William H., *Econometric Analysis*, 5th Edition, Upper Saddle River, N.J.: Prentice Hall, 2003.

Hanchate, A., N. McCall, and A. Ash, *The Adoption of Risk Adjustment Modeling for Prospective Payments*, Falls Church, Va.: TRICARE Management Activity, Health Program Analysis and Evaluation Division, Center for Health Care Management Studies, RTI Project Number 08490-013, June 2006.

Huang, I-Chan, Francesca Dominici, Constantine Frangakis, Gregory B. Diette, Cheryl L. Damberg, and Albert W. Wu, "Is Risk-Adjustor Selection More Important Than Statistical Approach for Provider Profiling? Asthma as an Example," *Medical Decision Making*, Vol. 25, No. 1, 2005, pp. 20–34.

Linden, Ariel, John L. Adams, and Nancy Roberts, "An Assessment of the Total Population Approach for Evaluating Disease Management Program Effectiveness," *Disease Management*, Vol. 6, No. 2, 2003, pp. 93–102.

Luft, Harold S., Sandra S. Hunt, and Susan C. Maerki, "The Volume-Outcome Relationship: Practice-Makes-Perfect or Selective Referral Patterns?" *Health Services Research*, Vol. 22, No. 2, 1987, pp. 157–182.

Marshall, Guillermo, A. Laurie W. Shroyer, Fred L. Grover, and Karl E. Hammermeister, "Time Series Monitors of Outcomes: A New Dimension for Measuring Quality," *Medical Care*, Vol. 36, No. 3, 1998, pp. 348–356.

McGlynn, Elizabeth A., Paul G. Shekelle, et al., *Identifying, Categorizing, and Evaluating Health Care Efficiency Measures*, Rockville, Md.: U.S. Department of Health and Human Services, Agency for Healthcare Research and Quality, 2008. As of June 25, 2009:
http://www.ahrq.gov/qual/efficiency/index.html

Military Health System, *MHS Strategic Plan Balanced Scorecard Overview*, no date. As of April 21, 2009:
http://www.ha.osd.mil/strat_plan/MHS%20Strategic%20Plan%20Overview.ppt

———, *Military Health System Strategic Plan*, April 2007. As of April 27, 2007:
http://www.ha.osd.mil/strat_plan/MHS_Strategic_Plan_07Apr.pdf

Newhouse, Joseph P., Melinda Beeuwkes Buntin, and John D. Chapman, "Risk Adjustment and Medicare: Taking a Closer Look," *Health Affairs*, Vol. 16, No. 5, 1997, pp. 26–43.

Office of Statewide Health Planning and Development (California), *Community-Acquired Pneumonia in California: Hospital Outcomes in 2002–2004*, Sacramento, Calif.: Office of Statewide Health Planning and Development, Healthcare Information Division, 2006.

Opsut, Robert, meeting with project monitor, December 20, 2006.

Solon, Jerry A., James J. Feeney, Sally H. Jones, Ruth D. Rigg, and Cecil G. Sheps, "Delineating Episodes of Medical Care," *American Journal of Public Health*, Vol. 57, No. 3, 1967, pp. 401–408.

Tapei Branch of Bureau of National Health Insurance, "Application of Catastrophic Illness Certificate," Web page, 2008. As of June 25, 2009:
http://www.nhitb.gov.tw/english/peopcure/illness.asp

TRICARE Management Activity—*See* U.S. Department of Defense, Office of the Assistant Secretary of Defense for Health Affairs, Health Program Analysis and Evaluation Directorate, TRICARE Management Activity.

U.S. Department of Defense, "Department of Defense Medicare Eligible Retiree Health Care Fund Operations," Instruction Number 6070.2, July 19, 2002. As of June 25, 2009:
http://www.dtic.mil/whs/directives/corres/pdf/607002p.pdf

U.S. Department of Defense, Office of the Assistant Secretary of Defense for Health Affairs, Health Program Analysis and Evaluation Directorate, TRICARE Management Activity, "TRICARE Costs," Web page, no date. As of June 25, 2009:
http://www.tricare.mil/tricarecost.cfm

——, *TRICARE Operations Manual 6010.51-M*, August 1, 2002. As of June 25, 2009:
http://manuals.tricare.osd.mil/

——, "Leveraging the Quadrennial Defense Review (QDR) to Transform the Military Health System," presentation at the State of the MHS—2006 Annual TRICARE Conference, Washington, D.C., January 31, 2006. As of June 25, 2009:
http://www.tricare.mil/conferences/2006/download/092Foster.ppt

——, *Evaluation of the TRICARE Program: FY 2007 Report to Congress*, 2007a. As of June 25, 2009:
http://www.tricare.mil/hpae/_docs/FY2007_2_27_07.pdf

——, "Prospective Payment System," presentation at the State of the MHS—2007 Annual TRICARE Conference, Washington, D.C., January 29, 2007b. As of June 25, 2009:
http://www.tricare.mil/conferences/2007/Mon/M400.ppt

——, "Per Member Per Month (PMPM): Metric Methodologies," presentation to the State of the MHS—2007 Annual TRICARE Conference, Washington, D.C., January 31, 2007c. As of June 25, 2009:
http://www.tricare.mil/conferences/2007/Wed/W420.ppt

——, "Transforming to Performance Based Financial Management," presentation to the State of the MHS—2007 Annual TRICARE Conference, Washington, D.C., January 31, 2007d. As of June 25, 2009:
http://www.tricare.mil/conferences/2007/Wed/W414.ppt

U.S. Department of Defense, Task Force on the Future of Military Health Care, *Final Report*, December 2007. As of June 25, 2009:
http://www.dodfuturehealthcare.net/

U.S. Department of Health and Human Services, Centers for Medicare and Medicaid Services, *Report to Congress: Plan to Implement a Medicare Hospital Value-Based Purchasing Program*, 2007. As of June 25, 2009:
http://www.cms.hhs.gov/AcuteInpatientPPS/downloads/
HospitalVBPPlanRTCFINALSUBMITTED2007.pdf

U.S. Government Accountability Office, *Medicare: Focus on Physician Practice Patterns Can Lead to Greater Efficiency*, Washington, D.C., GAO-07-307, 2007a. As of June 25, 2009:
http://www.gao.gov/cgi-bin/getrpt?GAO-07-307

———, *Military Health Care: TRICARE Cost-Sharing Proposals Would Help Offset Increasing Health Care Spending, but Projected Savings Are Likely Overestimated,* Washington, D.C., GAO-07-647, 2007b. As of June 25, 2009: http://www.gao.gov/new.items/d07647.pdf